D1994920

Luxembourg

An Emerging Cross-border Metropolitan Region

P.I.E. Peter Lang

Bruxelles · Bern · Berlin · Frankfurt am Main · New York · Oxford · Wien

Christophe SOHN (ed.)

Luxembourg

An Emerging Cross-border Metropolitan Region

Supported by the National Research Fund, Luxembourg (FNR/11/AM4/22).

Fonds National de la
Recherche Luxembourg

Cover picture: Gérard Borre, www.photonair.com.

No part of this book may be reproduced in any form, by print, photocopy, micro-film or any other means, without prior written permission from the publisher.
All rights reserved.

© P.I.E. PETER LANG S.A.

Éditions scientifiques internationales
Brussels, 2012
1 avenue Maurice, B-1050 Brussels, Belgium
www.peterlang.com ; info@peterlang.com

ISBN 978-90-5201-798-3
D/2012/5678/18

Printed in Germany

Library of Congress Cataloging-in-Publication Data.
Luxembourg : an emerging cross-border metropolitan region / Christophe Sohn (ed.).
p. cm. Includes bibliographical references and index.
ISBN 978-90-5201-798-3 (alk. paper)
1. Metropolitan areas--Economic aspects--Luxembourg. 2. Municipal government--Luxembourg--International cooperation. 3. Globalization--Economic aspects--Luxembourg. 4. Luxembourg--Economic integration. 5. Borderlands--Luxembourg. 6. Luxembourg--Boundaries. 7. Regional economics. I. Sohn, Christophe.
HT334.L8L89 2012 307.76'4094935--dc23 2012002320

CIP also available from the British Library, GB.

Bibliographic information published by "Die Deutsche Nationalbibliothek"
"Die Deutsche Nationalbibliothek" lists this publication in the "Deutsche Nationalbibliografie": detailed bibliographic data is available on the Internet at <http://dnb.de>.

Table of Contents

List of Tables

List of Maps

** See colour section.*

List of Figures

List of Abbreviations

BMVBS	Bundesministerium für Verkehr, Bau und Stadtentwicklung [German Ministry of Transport, Infrastructure and Urban Development]
CA2M	Communauté d'agglomération de Metz Métropole [Metropolitan Area of Metz]
CBD	Central Business District
CBPMR	Cross-border polycentric metropolitan region
CEO	Chief Executive Officer
CESGR	Conseil économique et social de la Grande Région [Economic and Social Committee for the Greater Region]
CES Lorraine	Conseil économique et social de Lorraine [Economic and Social Committee for Lorraine]
CPI	Conseil parlementaire interrégional [Interregional Parliamentary Council]
CSSF	Commission de surveillance du secteur financier [Committee on Financial Sector Supervision]
DATAR	Délégation interministérielle à l'Aménagement du territoire et à l'Attractivité régionale [French Interministerial Delegation for Territorial Planning and Regional Action]
ECSC	European Coal and Steel Community
EEC	European Economic Community
EGTC	European Grouping of Territorial Cooperation
ESDP	European Spatial Development Perspective
ESPON	European Spatial Planning Observation Network
EU	European Union
IFC	International financial centre
IGSS	Inspection générale de la sécurité sociale [General Inspectorate for Social Security]
INSEE	Institut national de la statistique et des études économiques [French National Institute of Statistics and Economic Studies]
IVL	Integratives Verkehrs- und Landesentwicklungskonzept [Integrated Transport and Spatial Development Concept]

M&A	Mergers and acquisitions
MERCOSUR	Mercado Comun del Sur [Southern Common Market]
MOT	Mission opérationnelle transfrontalière [Transfrontier Operational Mission]
NAFTA	North American Free Trade Agreement
OCSTAT	Office Cantonal de la Statistique [Cantonal Office of Statistics Geneva]
OECD	Organisation for Economic Co-operation and Development
PED	Pôle européen de développement [European Development Pole]
PFS	Professional of the financial sector
RFO	Regional Framework Operation.
SMOT	Schéma stratégique de mobilité transfrontalière [Transborder Mobility Strategic Scheme]
STATEC	Institut national de la statistique et des études économiques du Grand-Duché du Luxembourg [Luxembourgish National Institute of Statistics and Economic Studies]
UCI	Undertakings for collective investment

Acknowledgements

The idea for this book emerged at the end of the METROLUX (2007-2009) research project dedicated to the analysis of metropolitan issues and forms of governance in Luxembourg and its bordering territories.

This project – coordinated by the CEPS/INSTEAD public research centre, and with the support of the *Fonds National de la Recherche* (FNR) – specifically examined the establishment of a cross-border metropolitan region. Its aim was to analyse the economic dimension of this process, via the metropolitan attributes specific to Luxembourg; its spatial dimension linked to the cross-border scale of the regional integration process; and its political dimension resulting from the necessity of finding modes of governance capable of meeting the challenges of metropolisation within a borderland context.

Drawing on the results of the METROLUX project, the book also integrates work carried out in other research programmes, with the aim of offering a varied and coherent analysis of the different aspects of cross-border metropolisation in Luxembourg. Primarily derived from the investigations of geographers, this work combines conceptual reflection with empirical analysis. It constitutes a new contribution to the understanding of an atypical object of scholarship emerging from the geographic margins and the academic backstage.

The creation of such a work is, by definition, the result of joint collaboration. I would therefore like to sincerely thank first of all my colleagues in Luxembourg, France and Germany, for having agreed to act as authors for this project. This book is above all the fruit of their research. My thanks are also due to Emilie Menz, the editorial director at P.I.E. Peter Lang for her support and patience. Thanks also to FNR for the granting of financial assistance to support this publication and to Pierre Hausman, director of CEPS/INSTEAD, for having agreed to pay for the translation of this work. Finally, I would like to warmly thank the public authorities in Luxembourg and abroad who agreed to provide the data necessary for our research, as well as the many people who have dedicated their time to answering our questions and sharing with us their experiences and analyses.

Christophe Sohn

Luxembourg, July 2010

INTRODUCTION

The Concept of the Cross-border Metropolitan Region and its Relevance to Luxembourg

Christophe SOHN

The acceleration of the globalisation of economic and cultural exchange over recent decades has resulted in a profound restructuring of the geographical framework at the political and economic levels (Scott 1998). This spectacular spatial mutation has led to two distinct phenomena which are nonetheless interrelated.

Firstly, the concentration of people, resources and high-order services within large metropolises has been accompanied by a networking of these "global cities" to harness flows of labour, capital and information (Sassen 2001, Taylor 2004). Facing growth, international exchange and the arrival of a regime of flexible accumulation, the metropolises offer points of anchorage favourable to networks of commerce and communication. They constitute the new bases of the development of the information economy (Castells 1989), which are now organised as an archipelago (Veltz 1996). Such a spatial reorganisation of the economy and society leads to a redefinition of the traditional prerogatives of the state vis-à-vis the urban centres and a revival of cities – or more precisely metropolitan regions – as territorial actors (Brenner 2004, Le Galès and Harding 1998).

Secondly, the consolidation of regional integration mechanisms such as the EU and, to a lesser extent, NAFTA and MERCOSUR, has led to increased permeability of state borders in certain regions of the world. This relative opening up of borders has contributed to a larger process of territorial restructuring at different levels, of which the region is an essential one (Agnew 2000, Scott 2001). Such an increase in cross-border relations can be manifested in an intensification of economic exchange between bordering regions, the development of institutional cooperation or restructuring at the level of culture or identity (Perkmann and Sum 2002).

Despite the coexistence of the two phenomena – the close links they have with globalisation and the "new regionalism" that they promote –

they have different geographies and thus rarely confront each other. However, by taking the example of Luxembourg, which is characterised by the emergence of a cross-border metropolitan region, this study examines the intersection of these two dynamics. It aims to explain the connections between metropolisation and borders, as well as the characteristics of a spatial object that remains unrecognisable, despite being emblematic of the dialectical relations between the global and the local.

The first part of this introduction aims to specify the object of study by addressing the geographical realities that constitute it and critically examining the concepts used to understand it. While an attempt to summarise current thinking on the question remains incomplete, given the emergent character of the dynamics in question, it nonetheless allows for a better definition of the "cross-border metropolitan region". In the second part, these conceptual considerations are used to analyse the nature of the phenomena of cross-border metropolisation occurring in Luxembourg and in the bordering regions and to examine the characteristics of these. The final part sets out the structure of the book and examines the individual contributions.

Origin and development of the concept of the cross-border metropolitan region

Consideration of the issues surrounding cross-border metropolitan regions leads to reflection on the relations between the two objects that characterise the former: the city, in a broader sense, and the national border. As emphasised by the analyses of Saez, Leresche and Bassand (1997), as well as the work of Reitel *et al.* (2002), consideration of such a relationship is not only recent but in many ways counterintuitive. After presenting the foundations of this, this part attempts to describe the characteristics of the metropolitan phenomenon in a cross-border context. The analysis proceeds by means of an examination of the way in which the concept of the cross-border metropolitan region has been understood in academic literature. Finally, a definition of the concept is proposed.

Between city and border: a relationship that is not self-evident

As a component of the nation state, the border has long evoked the idea of periphery or outermost extremity, of closure and emptiness, while the city has always been associated with the idea of centrality (economic, political and cultural), of accumulation and connectedness. The approach to borders taken by regional studies, and economic geography referred to by Van Houtum (2000) as the "flow approach", is emblematic of this vision, with borders seen as barriers generating distortions in markets and border regions considered as economically disadvantaged areas that are not conducive to urban development. While from the point of view of political geography, the two objects bear the mark of state, their purposes differ: the border symbolises the limit of sovereignty while the city is part of the imposition of power (Reitel *et al.* 2002).

These classical understandings refer to a spatial reality in which the border zone is a marginal space from the point of view of economic and social development (see in particular Christaller 1933 and Lösch 1940). This situation is associated above all with the desire of states not to develop economic and social systems at the margins of their territory that are difficult to control and likely to generate covetousness (Saez and Bassand 1997). To this can be added the military imperatives that confer on borderlands the status of military buffer zones (Foucher 1991). These conditions, which accompanied the formation of modern states from the 16th century onwards, explain why the major cities and, *a fortiori*, the capital cities, are only rarely located close to an international border. While borders are generally not highly urbanised, two exceptions should be noted. First, there are cities that predate the establishment of a border and some of these have been able to benefit, legally or illicitly, from cross-border differentials and advantages they have been able to extract in order to develop and become relatively important urban centres (for example, Basel, Lille and Geneva). Second, there are border towns that have been created *de novo* by state decree. Numerous fortified locations have thus been built in order to strengthen the defensive function of a border and proclaim the power of a sovereign state (Denys 2002). In some cases – although these are rare – the creation of a border town following the imposition of a new border was done for economic or administrative reasons and not strictly military ones (such as Haparanda on the Sweden-Finland border). Whatever their origin, all the border cities remain places that are not only located close to a border but also are dependent on the border for their very existence (Buursink 2001).

New articulations resulting in a shift in perspective

Following the relativisation of the role of the state in economic and social regulation, and the exercise of political power (Jessop 2004), new perspectives have arisen regarding the relationship between city and border. The development of functional urban systems, though discontinuous, is no longer limited by national borders but rather increasingly concerns multi-national areas (Herzog 1990). This involves an effect linked to the opening up of borders and the intensification of economic and cultural relations. The border has changed from a "barrier" to an "interface" or "junction" (Strassoldo 1970, Courlet 1988, O'Dowd

2003) and the border regions have thus changed from being areas disadvantaged by their peripheral location to being potentially prosperous zones of socio-economic contact (Hansen 1977, Ratti and Reichman 1993, Martinez 1994, Clement 1997, van Geenhuizen and Ratti 2001). For the cities, this porosity of state borders – with regard to the circulation of goods, capital, services and people – constitutes an opportunity to take advantage of cross-border differentials (in particular in relation to tax and regulation) and to exploit the positive benefits that these represent for firms and workers. These changes occurred in different parts of the world over recent decades, but according to different time frames and modalities depending on the geopolitical and regional contexts. While the perspective arising from open borders is still particularly relevant in Europe, the post 9/11 rebordering process and the present border security regimes in North America have made it more difficult for stakeholders to take advantage of the borders and their differentials.

Within the EU, the rise of transborder urban spaces has led to cooperation that uses the border to strengthen the links between neighbouring countries and develop new modes of governance. Initially, these cross-border arrangements were conducted at the level of cross-border agglomerations and helped to resolve the local problems resulting from the proximity of the border such as trans-shipment within transportation systems or planning a joint infrastructure. It is only since the start of the 1990s that formal cross-border cooperation on a metropolitan scale emerged, in particular around European cities such as Basel (*Eurodistrict Trinational de Bâle*), Geneva (*Projet d'Agglomération franco-valdo-genevois*), Copenhaguen-Malmö (Øresund Committee) and Lille (Lille-Kortrijk-Tournai Eurometropolis). The development of these formal arrangements is linked to the promotion of legal tools and financial incentives dedicated to cross-border networking (e.g. INTERREG). Cross-border metropolitan cooperation involves not only a change of scale, but also a change of register as the strategic functions of metropolises such as economic competitiveness, the strengthening of the interconnections of major transport networks, territorial planning and international influence are concerned.

The existence of a metropolitan dimension to certain border regions has also been taken into account by the states in their regional development policies and strategic visions. In many cases, this has involved a complete reversal of perspective in that the national authorities responsible for spatial planning had long considered the national borders to be an insuperable limit. Thus, in France, the Interministerial Delegation for Territorial Planning and Regional Action (Délégation interministérielle à l'Aménagement du territoire et à l'Attractivité régionale, DATAR) has included eight cross-border metropolises (out of a total of 16 sites) within its call for metropolitan cooperation launched in 2004 and intended to consolidate the influence of French metropolises at the European level. In a similar way, the German Ministry of Transport, Infrastructure and Urban Development (*Bundesministerium für Verkehr, Bau und Stadtentwicklung*) has developed a planning strategy for the country that includes a section dedicated to metropolitan areas and promotes the development of cross-border metropolitan networks (BMVBS 2011). In Switzerland, the Federal Statistical Office recognises two cross-border metropolitan areas in its territory: Geneva-Lausanne and Basel. Last but not least, and relevant for this book, the Luxembourgish government and its regional partners within the Greater Region (Lorraine, Rhineland-Palatinate, Saarland and Wallonia) have also decided to promote the development of a cross-border polycentric metropolitan region (see Sohn and Walther, Chapter 11). To these state initiatives aiming to foster the economic competitiveness of these new peripheral centres, one should add the interest that central governments have in maintaining a certain level of control over their border regions undergoing the process of metropolisation.

The cross-border metropolitan regions are also subject to growing interest from the European bodies that see them as new economic and political units able to address specific issues in terms of governance. While the European Spatial Development Perspective (ESDP) does not specifically target cross-border metropolitan regions, the strategic orientation document does state that "co-operative cross-border city networks can provide a means of overcoming development disadvantages in border areas" (European Commission 1999: 21). More recently, the issue of cross-border metropolises has been addressed in a specific way within the framework of the EU's programmes promoting territorial cooperation. It must be noted that in relation to the objective of promoting a polycentric Europe, the cross-border metropolitan regions represent new zones of global economic integration able to counterbalance the economic hegemony of the "pentagon": the space between London, Hamburg, Munich, Milan and Paris, which includes almost 32% of the population and 43% of the GDP of Europe. For the European Spatial Planning Observation Network (ESPON), this interest is reflected in particular in the launch in 2009 of the METROBORDER project, which aims to improve the governance of polycentric metropolitan regions in Europe and in particular within the Upper Rhine and the Greater Region.

In summary, it should be emphasised that alongside economic and social interactions, political initiatives and their spatial manifestations, the emergence of cross-border metropolitan regions also takes place as for any spatial object as a discursive construction (Mondada 2000).

After this brief presentation of some elements of the discourse of the actors responsible for spatial development and its planning, a closer examination of the way in which cross-border metropolitan regions are understood within the academic discourse is necessary.

The place of cross-border metropolitan regions within academic discourses

An attempt to trace the emergence, development and place within the academic literature of a spatial concept leads to the delicate question of the relationships between a set of geographical configurations that share a certain number of joint characteristics, the way in which it is understood and the words used to refer to it. A comparative examination of these elements and their interaction allows some details to be discerned about the construction (which remains ever incomplete) of an object of knowledge. It is also necessary to use a body of publications representative of the richness of scientific work and remain aware that such an exercise must inevitably remain subjective.

A systematic bibliographic analysis based on SCOPUS data relating to the notion of the "cross-border metropolitan region" leads to an initial observation that the publications that understand this as a separate spatial category remain rare; there have been just 11 since 1990 (see figure I.1). Of these, attention should be given to the pioneering work of Herzog (1991) with its notion of the "cross-border metropolis". As a comparison, the older concept of the border city is present in 89 publications. However, articles concerning the process of the urbanisation of border zones and the resulting problems in terms of transportation, spatial planning and the environment are much more numerous (n=136). However, in the majority of cases (n=111), the cross-border context is associated with a generic understanding of cities, which covers highly disparate urban realities, from border towns to major metropolises. The limited number of studies on cross-border urban regions as a spatial object is reflected in the absence of a specific term. However, such a lexical variation must be interpreted with care. In certain cases, this reflects primarily the existence of differences in discursive practices within socio-cultural networks of thought (Bailly 1999). While the Europeans speak readily of the "metropolitan region", Anglophone researchers have generally preferred the term "city region", which, at this level of analysis, refers to the same thing. In other cases, differences of vocabulary reflect objectively distinct geographical realities, in particular in relation to the size of the cities in question. It is true that a small metropolis with several thousand inhabitants such as Basel, Geneva or Luxembourg must be distinguished from megacities such as Hong Kong, Singapore or San Diego-Tijuana, which exert influence over transnational sub-systems of several million inhabitants, despite the existence of a common denominator to these two categories of objects associated with the presence of national borders.

Figure I.1 – Bibliographic research into cross-border metropolitan regions in SCOPUS

> The bibliographic searches were based on articles and reviews published since 1970 in social and human science publications indexed by SCOPUS (over 5,300 titles). The results reflect the situation as of December 2009.
> The first analysis was based on the concept of the "cross-border metropolitan region" as a spatial object. The aim was thus to identify publications that make explicit use of this term or a set of terms using synonyms or related words. In the absence of the specific term, it was necessary to expand the searches to include all the expressions commonly used. Thus, the cross-border dimension was included via three terms: *cross-border*, *transborder* and *transfrontier*. In the same way, the concept of "metropolitan region" was addressed using five different terms: *city region*, *metropolitan area*, *metropolitan region*, *urban region* and *metropolis*.
> The second analysis aimed to document the existence of academic articles relating to a process of urbanisation or metropolisation of cross-border regions using the co-existence of the terms cited above within a single publication. In this case, it was not the "cross-border metropolitan region" as a spatial object being investigated but rather the consideration of a context.
> While the SCOPUS bibliographic database includes publications in various languages, the searches used only the English terms. As all references indexed in SCOPUS include abstracts and keywords in English, the linguistic bias is limited. Furthermore, in order to prioritise those publications that are most relevant to the issues examined, the searches used only titles, abstracts and keywords.

An examination of the geographical foundations of studies carried out shows a tendency towards diversification in terms of case studies. From 1980-90, the cases of Geneva (Switzerland-France) and Basel (Switzerland-Germany-France) in Europe, and El Paso-Ciudad Juarez (United States-Mexico) and San Diego-Tijuana (*ibid.*) in North America, primarily drew the attention of researchers (see Bailly 1987, Scott 1989, Gildersleeve 1978 and Herzog 1990). These are the most striking examples of cross-border economic integration: the concentration of capital industry and tertiary activities within urban border regions, accompanied by a growth in cross-border flows of workers, goods and information. In the case of the Swiss agglomerations, this economic and social interdependence gave rise to the development of cross-border alliances, which conferred to these areas the role of a laboratory for territorial and institutional restructuring at the margins of states (Leresche, Joye and Bassand 1995, Lotscher 1991, Jouve 1996). Over the following decade, the number and diversity of case studies increased, taking into account the reinforcement of cross-border metropolitan integration and the interest aroused by this phenomenon. In

Europe, the cross-border cooperation taking place in Copenhagen-Malmö (Denmark-Sweden), Lille (France-Belgium), Strasbourg (France-Germany) and Vienna-Bratislava (Austria-Slovakia) were the object of academic studies (see Hansen and Serin 2007, Baert 2008, Reitel 2006, 2007, Jasso 2007). In North America, the case of Detroit-Windsor (United States-Canada) was added to the potential cross-border metropolises located on the border between the US and Mexico (Brunet-Jailly 2000). Finally, in Asia, certain studies of the city-states of Hong Kong and Singapore also participated in this trend (Chen 2005, Fau 1999). For the former, the intensification of economic relations with mainland China, and the new perspectives in terms of cross-border cooperation in the Pearl River Delta, allow the emergence of a "cross-boundary global city region" to be observed (Shen 2004, Yang 2005, 2006). In the second case, the establishment of cooperation between Singapore, Johor (Malaysia) and the Riau archipelago (Indonesia) from 1989 onward gave birth to a cross-border economic region known as the "Sijori Growth Triangle" (Van Grunsven 1995).

In conclusion, it should be emphasised that the increase in the number and diversity of analyses relating to cross-border metropolitan regions certainly reflects a strengthening of transborder urban space integration in different parts of the world. The growth of these studies also reflects increased interest in the areas formerly considered as peripheral which, in accordance with changes resulting from globalisation and efforts to construct cross-border polities, are now seen in a new light. That said, given the variety of case studies and the concepts used to address them, the academic discourse appears to be somewhat confused, which ultimately poses the question of how clarification of the concepts used can be achieved.

The usefulness of defining the "cross-border metropolitan region"

Given a multifaceted object for which it is not easy to see the totality of possible configurations, nor the meaning that these should be given, it is debatable whether a definition can be elaborated. Would it not be better to leave the area of discussion open in order not to risk excluding certain phenomena of potential interest? Furthermore, how can one take into account the diversity of configurations and contexts in a definition that would necessarily render thought more rigid? While such arguments have good grounds, they are nevertheless strictly limited in their ability to place the object within a larger framework in order to analyse its specific characteristics. Indeed, offering a definition allows one to understand both that which is shared and that which proves different (Lévy 1997). In our approach, the differentiation relates to two different levels. First of all, there are differences between cross-border metropolitan regions and other urban configurations – in other words the specificity of the object and the interest in its study. Second, there are differences between case studies that lead us to investigate in the second part of this introduction the characteristics unique to Luxembourg as a cross-border metropolitan region.

In his pioneering work on San Diego-Tijuana, Herzog defines the cross-border metropolis as an "urbanized area fused into a single functional spatial domain that transcends the international border" (Herzog 1990: 139). This definition constitutes a good starting point. It takes into account the change of scale in the assessment of the urban space and adopts a dynamic understanding of the process at work. Such an approach appears, however, insufficient to evaluate the specific character of the complex object that is the cross-border metropolitan region. On the one hand, its descriptive dimension linked to a particular case study does not appear able to integrate the full diversity of existing spatial configurations. The geographical form of cross-border metropolitan regions can be subject to great variations as a function of monocentric or polycentric configurations of a region, the position of a border in relation to the metropolitan centre or centres and the scale of the urban system in question (metropolis or megalopolis), (see in particular Vandermotten 2010). On the other hand, this focus on form does not tell us much about the processes at work and their underlying mechanisms.

In order to explain the nature of the object, some additional considerations must be raised. On a global scale, a cross-border metropolitan region is a dense arrangement of knowledge, capital and labour, functionally integrated within the world economy and its transnational networks. Such global embeddedness is reflected by the concentration of metropolitan functions such as decision-making and control (regulation), innovation and competition, gateway and symbolic activities (Akademie für Raumforschung und Landesplanung 2007, Korcelli-Olejniczak 2007). On a regional scale, a process of cross-border integration is at play. According to Lee (2009), integration is defined as "the creation and maintenance of intense and diverse patterns of interaction and control between formerly more or less separate social spaces". Before moving forward, three clarifications must be made. First, the interactions that underlie cross-border regional integration are not limited to the economic sphere, but also concern other flows or transactions like migration, political trust, institutional ties, cultural exchanges (Väyrynen 2003). Second, the existence of such interaction across a border does not necessarily mean that the territories converge. Some relationships can be highly asymmetric and can be fed by strong differentials. If the possible convergence of the territories that interact across the border should be considered insofar as it raises the issue of their social and territorial cohesion, this dimension is not a prerequisite for defining a metropolitan region, whether domestic or cross-border. Third, the existence of strong functional interactions does not necessarily result in strong institutional integration (Sohn, Reitel and Walther 2009). Also, the creation of formal cross-border institutions or political alliances may take place despite relatively weak economic interdependencies or functional exchanges. As for any region, cross-border metropolitan regions are politically and socially constructed (Perkmann 2003).

However, what makes the object specific is the active role of the border in the coupling between the different scales of integration at play described above. The central hypothesis that underlines our thought is that the relative opening of borders represents new opportunities for border city-regions to reinforce their position into the global economy and strengthen their cooperative regional identity. While it is of course not a matter of minimising the barrier effects and the obstacles that borders can give rise to, it is necessary to recognise, following Hansen (1977, 1981), that the border context may also offers advantages. At the economic level, the exploitation of a border can strengthen the competitiveness of the metropolis via utilisation of differentials (especially in terms of salary costs and fiscal regulations) in a context marked by the relative porosity of borders and their increased permeability in terms of goods, capital, services, knowledge and people. The border can also represent a political resource for the actors in border regions engaged in the construction of territorial coalitions and alliances that transcend institutional divisions and offer them new perspectives. Finally, the resource can be symbolic in nature, when the border is used as an object of recognition in order to strengthen the international image of the metropolis and its influence (Reitel 2006, Sohn 2010). On the basis of these considerations, a cross-border metropolitan region can be defined as a region-level urban configuration inserted into the networks of globalisation and whose economic, cultural or political structuring is significantly linked to the existence of cross-border interactions.

Focusing on the case of Luxembourg and its surrounding territories, the analyses developed in this book will mainly address the economic and political dimensions of the process of cross-border metropolitan integration.

Luxembourg: a cross-border metropolitan region?

Seeking to analyse the phenomenon of cross-border metropolisation using the example of Luxembourg may lead to surprises. The approach thus requires several clarifications. With some 90,000 inhabitants, Luxembourg is not one of the major European cities. However, the city's international influence and economic attractiveness far exceed those that would be predicted on the basis of its limited number of inhabitants. Of the 180 agglomerations included in the comparative analysis of European cities, Luxembourg takes the 51st place (Rozenblat and Cicille 2003). In his analysis of global city networks based on relations between advanced producer service (APS) firms, Taylor (2000: 15) refers to "relatively strong evidence of world city formation" for Luxembourg.

First, we attempt to find out a little bit more about the origin of the metropolisation of Luxembourg and its specific characteristics. This initial investigation of the cross-border metropolitan region leads us to then question the relevance of the concept of metropolis to the case of Luxembourg. The final part will examine the role of borders within the dynamic of metropolisation and the resulting challenges.

A process of cross-border metropolisation driven by the state

The emergence of the metropolitan functions that allowed Luxembourg to insert itself into global and financial economic networks originated in the 1970s when the state, anxious to deal with the expected decline in the traditional steel industry, decided to attract financial institutions via attractive tax and regulations (Bauler 2001). In a post-industrial world increasingly subject to the vagaries of global commercial flows, it appeared to the country's rulers to be essential to develop service activities and gain access to external markets, and finance appeared an effective means of doing this. The success of this niche policy was reflected in economic growth greatly above the average of European countries and was accompanied by strong specialisation in financial services for businesses. In 2005 the financial sector alone accounted for 19% of domestic employment, 31% of tax intake and 38% of the added value (Deloitte 2006). As the main beneficiary of this concentration of the activities most susceptible to agglomeration, Luxembourg City became home to the majority of knowledge-intensive jobs. The economic attractiveness of the city centre is such that Luxembourg City is today one of the rare European urban centres to have a ratio of jobs to residents of over 1.5. As the national capital and the home of such European institutions as the Court of Justice, Court of Auditors and the European Investment Bank, the city can also boast a central political and symbolic role that gives it a certain status on the international stage (see Trausch 1994, 2003).

Centred on the urban agglomeration of Luxembourg, the metropolisation is above all a matter of the state and not of the capital city, which plays only a secondary role within this process (Sohn and Walther 2009). It is the state, on the strength of its sovereign power, that is at the origin of the development of the metropolitan functions. Financial services, insurance, logistics and, more recently, research and development, are strategic sectors that benefit from consistent political support from the central government. It is also the state that pilots urban planning measures intended to give the city both functional and symbolic metropolitan attributes. The development of the Kirchberg Plateau – which houses the European district, the Central Business District and, more recently, Luxembourg's major cultural infrastructural projects such as the Museum of Modern Art (MUDAM) and the Philharmonic concert hall – was planned and implemented by the state via an urban development fund. This influence of the state can also be seen in the development of the cultural influence of the capital, as shown by the organisation of the Luxembourg and Greater Region European Capital of Culture 2007 event. The initiative in giving this event a cross-border dimension by involving the Greater Region was taken by the Luxembourgish Prime Minister in order to strengthen the city's role as a cultural centre on the regional and European level (Sohn 2009a). This predominant role of the state within the process of metropolisation should be seen in the context of the territorial situation of Luxembourg. The small size of the country (2,586 km^2) and the proximity of the national and local scales serve to reinforce the state's capacity for action. In addition, the overlapping political nature of a bi-polar territorial structure (local and central governments) and the institutional fragmentation do not favour the autonomy of the country's 116 municipalities (Decoville 2008). In the absence of inter-municipal cooperation, the city, confined within its administrative limits, has only limited room for manoeuvre. The City of Luxembourg's status as national capital is accompanied, as is usual, by closer state supervision than applies to other cities.

An emerging "metropolis", or the predominance of process over form

Use of the term "metropolis" ("*métropole*") brings up first of all the delicate question of the definition of a "model which has not yet reached maturity" (Lévy 1995: 57). Etymologically, the term means the "mother city, founder of colonies." This meaning thus refers to the notion of a centre that controls and influences a large area. In practice, the term has long been used to refer to the main city of a country or region: an economic and administrative capital. This initial observation leads to two traditional meanings. The first relates to the functions of command and control concentrated in metropolises, which thus constitute their essential element. This concerns both decision-making functions relating to economic and cultural activities, reflected in the concentration of company headquarters and higher services, and the presence of a political and administrative authority such as a national or regional government. The second meaning, which follows from the first, emphasises the criterion of demographic size and is based on the observation that the larger the city, the greater its territorial influence will tend to be. The explanation relates to the fact that the advantages conferred by a metropolitan setting (size of local market, economies of urbanisation) lead to a significant concentration of the functions most susceptible to agglomeration in the largest cities (Fujita and Thisse 2002).

In light of these meanings, the case of Luxembourg appears at first glance to be problematic. While the growth of the financial centre and the presence of European institutions indicate a concentration of high value-added activities and command powers within the capital, this centralisation of metropolitan functions nonetheless fails to mask the fact that Luxembourg City does not have all the characteristics of a metropolis. The city does not have all the functions that are shared by the major metropolises, and the activities relating to culture and innovation appear to be behind, despite efforts in recent years to strengthen these strategic areas. In terms of demographic size, the agglomeration of Luxembourg, with its 130,000 inhabitants, is far from reaching international standards, with most authors agreeing that critical mass is reached at 500,000 to one million inhabitants (see Lacour and Puissant 1999). However, the size of a metropolis is necessarily relative to the dimensions of the country in which it is located (Pumain *et al.* 2006). Thus, qualification as a metropolis can vary as a function of the geographical and historical context within which an urban framework is found. Is it then necessary to consider Luxembourg as a special case?

On reflection, it is not so much the specificity of the case that raises a question but more the conceptual framework used to understand it. While the concept of metropolis remains associated with major urban agglomerations (from the mother city in ancient Greece to the large post-industrial city), the concept of metropolisation as a "contemporary form of the process of urbanisation" (Ascher 2003: 614) is much more fertile. It is ultimately a question of a process. Following Krätke (2007: 5), "metropolisation is a paraphrase for the increasing concentration of economic development potentials of the research-intensive industries and knowledge-intensive services on metropolitan regions and urban agglomerations". Over the course of globalisation, metropolisation does not restrict itself to the main urban centres but also concerns smaller urban entities (Julien 1995a). From this point of view, the case of Luxembourg underlines the limited interest of approaches that focus on population thresholds or the presence of activities and services that make a city into a metropolis, putting the emphasis instead on the economic, political, cultural and territorial dynamics that underlie the process of metropolisation.

To speak of metropolisation is above all to emphasise the ability of certain cities to adapt to the imperatives of globalisation and their insertion into globalised economic networks. It is also to emphasise the idea that being a metropolis is not a fixed and immutable characteristic but is rather subject to risk and possible alteration towards new forms of spatial structuration, on the basis of socio-economic, geopolitical or technological transformations. Given the incomplete nature of Luxembourg's metropolitan status, it is necessary to stress the still uncertain aspect of the dynamics at work. The very high specialisation of functions could represent a risk, as the durability of the activities driving growth is not certain over the medium and long term. Facing the uncertainties of a globalised economy, the metropolis is built up over the long term on the basis of feelings of belonging, shared values and reputation. Finally, mention of the term metropolisation is used to underline transformation of the forms and scales of urban spaces with the dilution of the urban areas beyond the built-up agglomeration contributing to the formation of huge functional ensembles (Ascher 1995).

Borders as resources for metropolitan development

In Luxembourg, metropolisation is accompanied by heavy functional integration of the German, Belgian and French bordering areas. Driven by the needs of a fast-growing financial sector, a heavy influence has been exerted over peripheral areas since the 1980s, in particular via the recruitment of cross-border workers. While there were less than 12,000 such workers in 1980, the number of employees from Lorraine (France), Rhineland-Palatinate or Sarreland (Germany) and Wallonia (Belgium) reached 146,000 in 2008 (STATEC 2009a). This cross-border mobility linked to employment is remarkable, both in absolute terms and in relation to the number of jobs available, as 44% of salaried positions in Luxembourg are occupied by cross-border workers. As a comparison, there where 52,600 cross-border workers in Geneva in 2008 and 40,000 between San Diego and Tijuana in 1998 (OCSTAT 2009, Bae 2003). When one examines the whole of the functional area of the agglomeration of Luxembourg defined on the basis of home-work commuters, the functional urban region includes almost a million inhabitants, which is almost double the population of the country (see Decoville and Sohn, Chapter 4).

In addition to the cross-border dimension of the metropolitan region, the decisive role of national borders within the dynamics of economic integration at work should also be underlined. The proximity of the borders allows Luxembourg to recruit skilled workers without having to bear the costs of their social reproduction. This strategy of exploiting the proximity of foreign labour is reinforced by a niche policy based on certain tax and regulatory protective measures associated with the sovereignty of the state. The cross-border differentials and the advantages that these generate are thus mobilised as an economic resource to promote the development of the metropolis. What is found here is undoubtedly a kind of "state metropolis" (CES Lorraine 2007), which takes advantage of its proximity to several national borders to strengthen its competitiveness and its integration within global production networks. Such a border regime is, however, not immune to potentially harmful effects, in particular the saturation of communication channels, which hinders the development of the employment area and the heavy functional specialisation of the space, which includes a metropolitan centre where activities and employment are concentrated and a border periphery relegated to the rank of dormitory towns. Given a niche policy, the limits of which can already be seen, crucial questions must be asked about the long-term future of the Luxembourg metropolitan region. What are the potentials and the threats to the metropolitan development of Luxembourg? What is the role for political support of a process in which economic dynamics appear to play the primary role? How are the functional interdependencies shaping the spatial configuration of the developing cross-border metropolis? How are issues relating to borders – up until now dominated by the imperatives of a small state that is nonetheless sovereign – taken into account in cross-border initiatives and institutional and partnership constellations? To what extent is a metropolitan construction subject to change, and thus how durable is it? While these issues raise the question of the genesis of the Luxembourgish metropolis, they also raise an essential question about its future.

Plan and contributions

The analysis of the emergence of a cross-border metropolitan region in Luxembourg offered in this book is based around three fundamental aspects. In the first section, entitled "Metropolitan centralities", the primary focus is on the economic and symbolic attributes that characterise the process of metropolisation and reveal some of its potentials and limits. The second part, "Cross-border interdependencies" examines the cross-border dimension of the dynamics of metropolisation at work, principally in terms of home-work mobility and residential practices. Finally, the third part, entitled "Governance stakes and strategies", analyses the political construction of a cross-border metropolitan region and the strategy of the actors involved in its governance.

Metropolitan centralities

As a process that concentrates command functions and strategic activities within cities positioned within the networks of the globalised economy, metropolisation is undoubtedly part of a process of affirming major urban centralities. Understood broadly as "the capacity of a place to exert influence over a greater or smaller area" (Dematteis, 2003), the concept of centrality applied in this first part refers to the characteristics of an urban centre and not to the flows and connectivity that these generate. From this territorial perspective, two aspects of metropolitan centralities are privileged. In order to better grasp the characteristics of the metropolisation of Luxembourg and the processes that underlie its development, Chapters 1 and 2 focus on the economic attributes of the metropolis. Meanwhile, Chapter 3 examines the international influence of the metropolitan centre on the business world. Whatever analytical approach they take, these contributions all aim to better understand the advantages, specific characteristics and limits of an emerging metropolis.

In Chapter 1, Olivier Walther examines the profile of the Luxembourg economy and its development over the last 15 years by focusing on knowledge-intensive employment. Adapting the Eurostat/OECD nomenclature for the case of Luxembourg, the study underlines the growing importance of so-called "metropolitan" jobs to Luxembourg and the heavy specialisation of the country as a financial centre and home to European institutions. On the geographical level, Walther discusses the remarkable spatial concentration of knowledge-intensive activities within the Luxembourg agglomeration, which contributes to strengthening the capital's position as a metropolitan centre.

Limiting the field of study to the key activity driving metropolisation in Luxembourg, in Chapter 2 Olivier Walther and Christian Schulz examine the specific development of the financial centre. By studying the effects exerted by the forces of concentration and dispersion on the financial cluster, this chapter underlines the advantages and disadvantages of Luxembourg City as a privileged site for the financial industry in order to question its long-term sustainability. Less and less dependent on a niche policy with an uncertain future, the development of the sector appears to be now increasingly linked to the presence of externalities intrinsic to a metropolitan environment, in particular the diversity and quality of services offered to businesses. To this should be added the presence of an advantageous institutional configuration resulting from the small size of the country and the central role played by public actors, in particular the state, in the metropolitan development of the country.

Finally, Chapter 3 offers a different perspective: Ralf Bläser, Michaela Gensheimer and Christian Schulz examine the way in which Luxembourg's metropolitan status is perceived in the sphere of international business. This analysis, which looks at the views of financial sector managers and the portrayal of the capital in the international press, draws attention to the discursive construction of the Luxembourgish metropolis. The metropolisation is not based solely on the concentration of knowledge-intensive economic activities, but also mobilises symbolic resources, images and collective representations that undoubtedly contribute to strengthening the attractiveness of the urban centre, but also grant the city a social significance.

Cross-border interdependencies

The second part of the book expands the scale of analysis to examine the cross-border relations linking Luxembourg and the neighbouring German, Belgian and French territories that constitute the Greater Region. Using the concept of interdependence, this section aims to emphasise the importance of socio-economic interactions that link the populations and territories on either side of the national borders. As well as the peripheral border areas being dependent on the Luxembourg metropolitan centre, one can also observe that the Luxembourg economy is dependent on cross-border labour, which represents almost a third of salaried positions within the country. This situation is not without consequences in terms of the spatial and social dynamics at work within this resolutely cross-border area of metropolitan integration. While Chapter 4 examines the geography of the functional urban region, Chapters 5 and 6 focus on the daily and residential mobility practices and the factors that underlie them. Chapter 7 concludes the section by examining the demographic development of the border areas in order to test a hypothesis of territorial convergence.

In Chapter 4, Antoine Decoville and Christophe Sohn consider the spatial form of the functional urban region that is developing under the influence of the home-work mobility influenced by the capital city. Arguing that it is misleading to seek to give a single unique definition of the functional urban region of Luxembourg, the authors use different cartographic representations of a single phenomenon in order to better grasp a reality that is fluid and multifaceted. While it is clear that the employment area of Luxembourg is tending to become more and more cross-border in nature, the dynamic of spatial expansion of this functional urban region appears to be running out of steam and is being replaced by a strengthening of the border urban centres that are well connected to major roads. However, these socio-spatial restructurings are subject to different residential practices as a function of the sectors in which the commuters work.

In order to understand in greater detail the mechanisms and spatial consequences of this specialisation, both functional and social, Philippe Gerber, Olivier Klein and Samuel Carpentier analyse in Chapter 5 the system of local mobility. Addressing both practices of home-work mobility and the residential choices of cross-border and resident workers, this chapter illustrates the functional interdependencies that develop within peripheral, national and bordering areas subject to a process of peri-urbanisation. The analysis also underlines the key role of national borders within the geography of local mobility flows, either as catalysts of cross-border working and thus home-work mobility, or as attractors, coupled with a barrier effect, in relation to residential trajectories.

In Chapter 6, Sébastien Lord and Philippe Gerber offer a detailed examination of the social dimension of residential mobilities, particularly in relation to the manifestation of socio-spatial inequalities within the cross-border employment area. Their study concerns the socio-demographic characteristics of the individuals who relocate and aims to explain the factors influencing residential mobility in Luxembourg and the bordering regions. Among the multiplicity of possible trajectories, particular attention is paid to residential relocations that involve: 1) a move closer to the Luxembourgish border; 2) a move across the border into Luxembourg or the capital; 3) a move out of the country into a neighbouring country; and 4) relocations that do not involve the border. Statistical modelling confirms that workers in metropolitan employment can more easily cross the border to move closer to the metropolitan centre, although this dynamic of socio-residential segregation at the top of the social scale is mitigated by the effect of other factors determining residential trajectories, such as life cycles and attachment to one's region of origin.

While the above contributions are based on the analysis of the functional integration of a cross-border employment area, taking into account the spatial practices of commuters, Chapter 7 concludes this section by adopting a different and complementary perspective. Examination of population densities and their development allows Olivier Walther to determine whether or not the presence of strong cross-border interaction results in the convergence of the demographic characteristics of the bordering territories. The author examines not only the demographic integration within the Luxembourg metropolitan area, but also in Basel and Geneva. While the three cases studied are metropolitan regions heavily integrated from the point of view of the employment market, the analysis of their demographic dynamics provides divergent results. While for Basel and Geneva a certain degree of demographic convergence of the bordering districts can be observed, the case of Luxembourg is marked by the presence of strong growth differentials across the borders, suggesting that the cross-border metropolitan territory remains morphologically fragmented.

Governance stakes and strategies

After analysing Luxembourg's position in metropolitan terms, and the functional interdependencies that the process of metropolisation creates with the bordering territories, the third part examines the governance of this emerging cross-border region. With different meanings depending on the context, the concept of governance is here understood as involving above all the existence of an increasing multiplicity of actors and interests engaged in the political management of territories, in a context of economic globalisation, European integration and the questioning of traditional state prerogatives. Associated with the concepts of metropolis and border, cross-border metropolitan governance concerns the modes of coordination and regulation used within the framework of the construction of urban regions intersected by national borders. In order to identify the strategies of actors and their projects, and also the reasoning behind the partnerships and networks of actors found within the different arenas of cross-border cooperation, the following chapters use multiple points of view and scales of observation. Examining the Greater Region (Chapter 8), cross-border city networks (Chapter 9), the national spatial planning policy (Chapter 10) and the project of a polycentric metropolis (Chapter 11), this section questions the modalities of intervention of cross-border metropolitan governance.

In Chapter 8, Christian Lamour and Franz Clément examine the role played by the institutions within the Euroregion in the organisation of cross-border metropolitan development. Is there a Greater Region orientation in matters of metropolitan coordination? If so, in what form is this coordination manifested? This chapter's hypothesis is that the Greater Region contributes to the development of a multi-level network of public actors involved in the development of cities. To support their analysis, the authors outline the institutional development of the Greater Region, its territorial enlargement, the consolidation of its mechanism of governance and its openness to economic networks and civil society. They also show how the issue of cross-border metropolitan governance, which has been developing for many years within networks of expertise and cooperation, has progressively gained in importance in the political agenda of the Greater Region.

Following on from the preceding analysis, Chapter 9 seeks to examine the role of city networks in cross-border territorial governance. To do this, Eric Auburtin first discusses the place of the concept of polycentrism in the regional development of Europe and its transfer to the context of the Greater Region. On the basis of the observation that only a cross-border polycentric organisation appears capable of compensating for the absence of a true metropolis of European rank, the author emphasises the paradox of a greater region that to date has been largely constructed without the cities. The chapter then documents the role of actors within these alliances between urban centres and analyses the complex relations that these networks maintain with the territorial structures within which they are located. With their advantages of flexibility and pragmatism, the city networks are ultimately seen as embryonic forms of developing cross-border metropolitan governance, but whose effectiveness remains open to question due to the competition between the cities and the lack of integration of the interstitial spaces into the overall dynamic.

In Chapter 10, Antoine Decoville focuses on Luxembourg and seeks to decipher the territorial development policy at work, and in particular the role of central government. In contrast to the neo-liberal trend observed from the 1980s onward in the majority of European countries – which is reflected in territorial development policies by a reduction in the role of the state in the determination of spatial development and increased importance being granted to the imperatives of territorial competitiveness – the Luxembourgish state continues to operate an active policy of regulation of spatial dynamics. In order to examine in detail such an approach, this chapter focuses on the development of Esch-Belval, a former post-industrial steel-working area located in the south of the country and destined to house the future *Cité des Sciences* around a "new town" planned and financed by the state in partnership with private investors. With a combination of classical theoretical references, and innovative and pragmatic modes of operation, the political management of the urban development dynamics of Luxembourg is ultimately seen as the fruit of a state strategy that aims to support the growth of the Luxembourg metropolis and to position itself as a privileged partner for business.

In the final chapter of this section concerning governance, Christophe Sohn and Olivier Walther seek to clarify the political issues at stake in the construction of a cross-border polycentric metropolitan region at the heart of the Greater Region. Taking a constructivist perspective, Chapter 11 seeks to show that the strategies of cooperation implemented by the actors are in part conditioned by the role played by the borders within the political economy of the cross-border region, and in particular the fact that they are able to offer certain economic or symbolic advantages. This interpretation of the border-related issues allows the authors to then explain that in Luxembourg, the degree of functional integration, which is manifested in particular in the high number of cross-border workers, is not accompanied by cross-border cooperation on an equivalent scale, given the strategies of exploitation of cross-border differentials by the Luxembourgish state. This chapter concludes with an analysis of the current limits to such a situation and a critical examination of the stakes and perspectives that underlie the recent launch of cross-border cooperation on a metropolitan scale within the framework of the Greater Region.

The conclusion, written by Bernard Reitel, returns to the key aspects of Luxembourg's affirmation of its position as an emerging metropolis in order to draw up a balance of the issues with which this project, promoted by the state, is confronted. From this perspective, the author examines the conditions required for the metropolitan dynamic to be sustainable in the long term, the methods of the development of the functional region and the modes of governance able to transform what is still largely perceived as a "state metropolis" into a true cross-border metropolis. The book ends with a reflection on the features unique to the case of Luxembourg in the international context and its potential to serve as a model.

PART 1

METROPOLITAN CENTRALITIES

CHAPTER 1

Regional Growth of the Knowledge Economy in Luxembourg

Olivier WALTHER

Contrary to those theories that anticipated the end or decline of cities following the development of the new information and communication technologies, metropolitan regions are today considered the engines of economic development (Hall and Pain 2006). In these metropolitan areas that provide easy access to national and international markets, a major concentration of activity in the fields of knowledge-intensive services and industries has been seen (van Winden, van den Berg and Pol 2007, Krätke 2007). From the economic point of view, the success of these activities is based on a high density and large diversity of high-order service companies working at a global level (Taylor 2004) and of public and private-sector actors engaged in innovation processes.

In Europe, the concentration of these activities has been studied primarily in the largest metropolitan centres, as this is where the density of employment and the diversity of functions has been the greatest, whether in the case of London (Taylor *et al.* 2003) or Paris (Boiteux-Orain and Guillain 2004, Halbert 2004), or second-rank urban centres such as Munich (Lüthi, Thierstein and Goebel 2008), Zurich (Thierstein *et al.* 2008), Stockholm (Hermelin 2007) or Dublin (Sokol and van Egeraat 2005). The case of smaller cities, however, remains less well documented, although these can also contribute to forming highly qualified labour pools when they experience, as is the case for Luxembourg, a high level of specialisation in certain knowledge-intensive activities.

In this context, and on the basis of the work of the OECD (2006a) and Eurostat (2006) on the knowledge economy, this chapter has a double purpose. From the sectoral point of view, the aim is to propose a classification of economic activities seen as representative of the knowledge economy, adapted to the specific characteristics of Luxembourg, and to discuss the sectoral development of the jobs linked to these activities between 1994 and 2008. This analysis highlights the major trends affecting the economic profile of the Luxembourg metropolitan area, and confirms the growing importance of knowledge-intensive activities. Then, from the geographical point of view, the classification set out in this chapter allows the mapping of jobs characteristic of the knowledge economy and thus to identify the focal points of the metropolitan economy in space.

Agglomeration, producer services and the knowledge economy

The location of economic activities has traditionally been interpreted as being the outcome of two competing sets of forces: centripetal forces favouring concentration and centrifugal forces favouring dispersion (Krugman 1991). In the context of the concentration of economic activity at the global level, a very substantial literature has demonstrated the importance of classical Marshallian externalities, particularly the fundamental role of a specialised and/or skilled labour market that allows the recruitment of employees in a flexible manner and at low cost; the availability of specialist suppliers and clients, supplying products adapted to the market; and the impact of the introduction of technological spillovers, which allow benefits to be gained from the innovations introduced by companies (see Gordon and McCann 2000 for a survey).

Apart from these classical externalities, economic activities – and in particular high-order services – are also susceptible to three other types of factors (Porteous 1999). The first concerns informational spillovers that result from the proximity to the sources of information. Following the pioneering work of Polanyi (1967), it is generally acknowledged that the new information and communication technologies heavily favour the exchange of codifiable knowledge, while the exchange of non-codifiable knowledge or expertise generally requires physical proximity. The advantages that flow from these face-to-face exchanges have been recognised in the economic literature, in particular with regard to financial services, culture, politics and the development of new technologies (Boiteux-Orain and Guillain 2004, Storper and Venables 2004). These spillover effects are then supplemented by socio-cultural factors. Indeed, companies are socially embedded in a complex of norms and rules that are reinforced by proximity and that remain informal or gradually become institutionalised. The importance of this factor has been demonstrated in particular in the case of the financial services that develop within the City in a social context, and that favour trust between the actors and provides them with a reputation (Thrift 1994, McDowell 1997). Finally, it is undeniable that the agglomeration of economic activities can be favoured by the policies enacted to favour their implantation, particularly a competitive legislative framework. The example of Luxembourg is highly revealing from this point of view, as it shows that a financial industry initially based on a territorial niche policy can go on to generate significant agglomeration economies as the number and diversity of the actors within the sector in question increases (Pierreti and Bourgain 2003, Walther and Schulz Chapter 2).

These externalities thus lead to a significant concentration of the activities most susceptible to agglomeration, even though certain technologies increasingly allow distance to be overcome (Hall and Pain 2006). This paradox implies that there is simultaneous dispersal of those economic implementation activities most sensitive to flexibility and cost constraints (such as low-technology activities and certain services that can be easily relocated) and concentration of economic innovation and command activities sensitive to proximity (such as high-technology activities and certain knowledge-intensive services), (Sassen 2001).

These latter activities today form the core of the knowledge economy, defined as the ensemble of the "industrial activity branches with a high share of research and development activities for the generation of new technological knowledge as well as those economic activities for which the generation and economic use of specific knowledge is a main focus" (Krätke 2007: 4). It is generally recognised that this knowledge economy consists of two particular categories of activity: advanced producer services (APSs) that provide "any service functions that are ordinarily consumed by firms and organizations in the course of their operations" (Coffey 2000: 171) and high-technology industries, which are the source of pharmaceutical, chemical, medical and biotechnological products. Following Greenfield (1966), several generations of work have progressively documented the importance of producer services within the contemporary economy, particularly their export character and multiplicative effect on revenues and employment, as well as the way in which these activities can be subjected to different sectoral classifications (see Martinelli 1992). Three main approaches can be distinguished in the contemporary literature.

In *The Global City*, Sassen (2001) uses the existence of APSs to develop a profile of global cities and set out their role within a globalised economy. These services are aimed at all types of organisation (industrial or service enterprises, in the private or public sectors) and cover financial, legal and management services offered to companies and the insurance, innovation, communication, personnel management sectors, as well as those linked to transport, storage and security. Finance, insurance and real estate form a key sector designated by the acronym FIRE, together with management, accountancy, advertising, communication and legal services. One of the methodological difficulties presented by this kind of analysis is the fact that these activities are not provided solely to enterprises but also to individuals, and the distinction is not always easy to observe from a statistical perspective. In addition, the list of sectors of activities involved in services to enterprises varies from country to country as a function of national classifications.

In order to refine the statistical approach to the key economic functions of the metropolitan regions, other work has focused on the construction of the classifications adopted in national contexts. In France, for example, the National Institute of Statistics and Economic Studies (Institut national de la statistique et des études économiques, INSEE) has produced a definition of "high-order metropolitan functions" (Julien 1995b). This functional analysis of employment, which transcends analyses by sector or profession, was developed on the basis of the 1990 population census to complement the traditional analysis by sectors of activity and show the characteristic functions of large cities: those "for which the decisional content is increased or which contribute to the brand image of the city where they are carried out" (Julien 2002: 4). Constructed in an empirical manner, the list of high-order metropolitan functions is based on the intersection of economic activity and socio-professional category. Three criteria are utilised: the functions are easily identifiable; they are present in major cities; and they correspond to the highest level of qualification (management or engineer). The sectors identified include art, banking/insurance, commerce, the commercial sector within industry, management, media, IT, research, services to enterprises, telecommunications and transport. Although this classification of high-order metropolitan functions rests on solid empirical bases, it is not intended to be exhaustive and remains centred methodologically on the case of French cities.

Developed within the framework of the theory of knowledge and on a European level, the work initiated by the OECD in the mid-1980s is not subject to the above criticisms (OECD 2006a). Based on the intensity of research and development (R&D) and the technological level of activities, these analyses initially concerned the manufacturing sector (Hatzichronoglou 1997). Subsequently, they were extended by Eurostat (2006) to cover service activities, and finally provided a European classification of high-technology and knowledge-intensive sectors. The resultant classification distinguishes between four categories of the manufacturing industry as a function of their technological level, as well as six categories of services of which four are highly knowledge-intensive and two are less knowledge-intensive. Table 1.1 shows these different sectors on the basis of the NACE classification of activities adopted by the European Union (EU). The high-technology and knowledge-intensive services (KIS) discussed in the subsequent analysis appear in bold.

The main drawback of this sectoral classification is associated with the heterogeneity of the activities considered as highly or less highly knowledge-intensive. The financial sector, for example, consists of several very diverse specialisations (asset managers, IT workers, secretaries, security staff). Enterprises are also increasingly inclined to specialise in terms of type (function) of job and no longer only in terms of economic sector. The management and production functions thus tend to occupy different places within the same economic sector, as shown by Duranton and Puga (2005), who describe this as the passage from sectoral specialisation to functional specialisation.

Table 1.1 – Manufacturing and services sectors

Economic sectors	NACE version 1.1 codes
Manufacturing industry	
High technology	**24.4, 30, 32, 33, 35.3**
Medium-high technology	**24 (-24.4), 29, 31, 34, 35 (-35.1 and 35.3)**
Medium-low technology	23, 25 to 28, 35.1
Low technology	15 to 22, 36, 37
Services	
High-technology Knowledge Intensive Services (KIS)	**64, 72, 73**
Market Knowledge Intensive Services (KIS)	**61, 62, 70, 71, 74**
Financial Knowledge Intensive Services (KIS)	**65 to 67**
Other Knowledge Intensive Services (KIS)	**80, 85, 92**
Market Less Knowledge-Intensive Services (LKIS)	50 to 52, 55, 60, 63
Other Less Knowledge-Intensive Services (LKIS)	75, 90, 91, 93, 95, 99

Sources: OECD (2006a) and Eurostat (2006).

Note: high-tech industries and knowledge-intensive services are indicated in bold.

A sectoral classification adapted to Luxembourg

Justification of choice of data

The analysis of employment linked to the knowledge economy is conducted primarily on the basis of the data provided by the Luxembourg General Inspectorate of Social Security (Inspection générale de la sécurité sociale, IGSS). In contrast to the data from the population census which concerned only residents, the IGSS data have the advantage of taking into account both employees resident in Luxembourg and those who cross the border daily, as well as all their co-insured, which is advantageous given that 41.5% of jobs were held by cross-border workers in 2008 (STATEC 2009b). The data used thus relate to all those in active employment covered by social insurance provided by the state of Luxembourg on March 31 of each of the years for which data are provided (1994, 1996, 1999, 2002, 2005, 2008).

The IGSS data calculates employment on the basis of the head office of the enterprise and not of the employee's real place of work. For 2005, this statistical bias was corrected by the Luxembourgish National Institute of Statistics and Economic Studies (Institut national de la statistique et des études économiques du Grand-Duché du Luxembourg, STATEC) at the level of the principal national sections and divisions of the NACELUX classification of activities. Even if this detail has no impact on employment calculated at the national level, Figure 1.1 shows that significant differences can be seen at the level of the municipalities.

Figure 1.1 – Employment at head office and place of work in Luxembourg and Esch-sur-Alzette, 2005

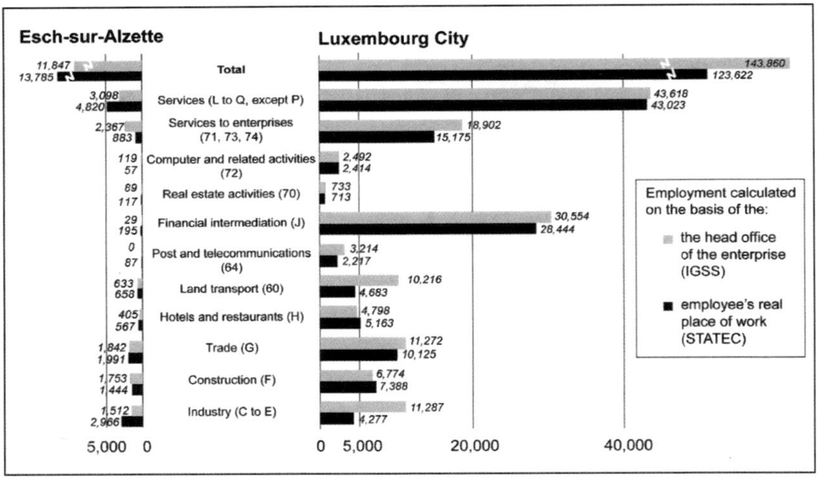

Sources: IGSS and STATEC.

Note: the sections and divisions from the NACE revision 1.1 classification used by STATEC are shown in brackets.

It can be seen that the number of employees recorded by the IGSS in Luxembourg City at the head office of enterprises (143,860) is significantly higher than the corrected figure given by STATEC for 2005 (123,622). Consequently, the proportion of jobs connected to the knowledge economy, calculated subsequently in the present chapter, is overestimated for Luxembourg City by location of head office (50.1%) in comparison to the calculations carried out on the basis of the real place of work (46.2%) for 2005. A more detailed comparison of the figures provided by the IGSS and STATEC shows that it is primarily the industry sections (+7,010) and the transport divisions (+5,533), together with services to enterprises (+3,727), that are over-represented in relation to real employment. Taking into account this methodological bias, which results in over-representation of employment at head offices, generally located in Luxembourg City, the following calculations have been made on the basis of two distinct sets of data. The IGSS figures based on head offices are used to analyse the growth of employment at the level of Luxembourg as they provide five-figure NACE codes and allow a more precise breakdown of employment than the divisions and sections provided by STATEC. This data, available from 1994, also allows a diachronic analysis of the development of knowledge-intensive employment. Otherwise, the corrected STATEC figures using real place of employment are used to give an indication of the precise location of enterprises connected to the knowledge economy at the level of Luxembourg City, which was the object of a detailed analysis in 2005.

Adaptation for the case of Luxembourg and identification of high-technology and knowledge-intensive sectors

The methodological process includes two stages. First, the OECD-Eurostat classification is adapted to the specifics of Luxembourg, to the extent to which this is mandated by the nature of the data used. Then, a procedure selecting the jobs corresponding to knowledge-intensive sectors is carried out.

The classification of knowledge-intensive activities produced by the OECD and Eurostat is based on the classification of economic activities within the EU (NACE version 1.1). Taking into account the data available at the European level, the identification of high-technology manufacturing industries has been made using the NACE codes to three figures, while the knowledge-intensive services are aggregated on the basis of the NACE codes taken to two figures. The use of figures from the IGSS's administrative files allows the classification to be refined and to better take into account the specific features of the Luxembourg economy by using the NACE code to five positions for all those in active employment and registered for social security. This statistical opportunity allows two types of modification to be made to the OECD-Eurostat classification (Table 1.2).

Table 1.2 – High-technology manufacturing and knowledge-intensive services sectors

OECD classification	NACE	METROLUX classification	NACE
Manufacturing industry		**Manufacturing industry**	
High-technology		**High-technology**	
Aerospace	35.3	Aerospace	35.3
Computers, office machinery	30	Computers, office machinery	30
Electronics-communication	32	Electronics-communication	32
Pharmaceuticals	24.4	Pharmaceuticals	24.4
Scientific instruments	33	Scientific instruments	33
Knowledge-intensive services		**Knowledge-intensive services**	
Knowledge-intensive high-tech services		**Knowledge-intensive high-tech services**	
Post and telecommunications	64	Telecommunications	64.2
Computer and related activities	72	Computer and related activities	72
Research and development	73	Research and development	73
Knowledge-intensive market services		**Knowledge-intensive market services**	
Water transport	61	Water transport	61
Air transport	62	Air transport	62
		Other supporting transport activities	63.2
		Activities of other transport agencies	63.4
Real estate activities	70	Real estate activities	70
Renting of machinery and equipment without operator and of personal and household goods	71	Renting of machinery and equip-ment without operator and of personal and household goods	71
Other business activities	74	Other business activities	74.1-74.5
Knowledge-intensive financial services		**Knowledge-intensive financial services**	
Financial intermediation	65	Financial intermediation	65
Insurance and pension funding	66	Insurance and pension funding	66
Activities auxiliary to financial intermediation	67	Activities auxiliary to financial intermediation	67
Other knowledge-intensive services		**Other knowledge-intensive services**	
Education	80	Higher education	80.3
Health and social work	85	Human health activities	85.1
Recreational, cultural and sporting activities	92	Recreational, cultural and sporting activities	92
		Other less-knowledge-intensive services	
		General (overall) public service activities, Foreign affairs, Justice and judicial activities (selection)	75.111, 75.210, 75.230
		Extra-territorial organisations and bodies	99

Sources: OECD (2006a) and Eurostat (2006).

Note: sectors disaggregated are indicated in light grey; sectors added to the classification are indicated in dark grey.

Firstly, certain sectors that are aggregated when the NACE codes are taken to two positions are disaggregated to select the most relevant activities (in light grey in Table 1.2). This operation was carried out for the following activities: post and telecommunications (64), other business activities (74), education (80) and health and social work (85). In each of these activities, only the sectors with the highest level of knowledge intensity as identified by Krätke (2007) in his analysis of the knowledge economy at the level of the European metropolises have been retained: telecommunications (64.2); certain other business activities (75.1-5); higher education (80.3); and the human health activities (85.1).

Secondly, the classification has been completed by sectors of activity particularly representative of the economic structure in Luxembourg (in dark grey in Table 1.2). This concerns the activities linked to the other supporting transport activities (63.2) and the organisation of freight transport (63.4), which are particularly developed thanks in particular to the Luxembourg airport facilities, as well as certain activities within the general (overall) public service activities, foreign affairs, justice and judicial activities (75.111, 75.210, 75.230) as well as the extra-territorial organisations and bodies (99).

This modified classification has been used as the basis for the identification of high-technology and knowledge-intensive jobs. Socio-professional status has been taken into account in order to exclude the analysis of blue-collar workers, as this category of employees generally corresponds to less qualified positions which are thus not representative of high-technology and knowledge-intensive employment. Although the IGSS data provides neither the NACE code nor the location of the head office of self-employed intellectual workers, this category has been taken into account in calculating high-technology and knowledge-intensive jobs because of the specific profile of the professions in question (doctors, lawyers, insurance agents, etc.). The international civil servants included by STATEC and absent from the IGSS data have also been included in the calculation of these jobs. These two additions, however, concern only the statistics aggregated at the national level, as no information relating to the location of these jobs is available for the dates studied.

Validation of methodological choices

Classically, the spatial distribution of services between cities is dependent on the nature of economic activity. Within public services or household services, which are distributed by the state as a function of the resident population for the former and as a function of local demand for the second, the number of jobs is generally inversely proportional to the rank of cities within an urban system. For high-order services, in contrast, the number of jobs is not statistically connected to the demographic size of the cities. These jobs are disproportionately more numerous in those cities of superior rank than in other centres within the urban system (DATAR 2004). This particularity is due to the economies of agglomeration, which favour the concentration of high-order service enterprises within metropolises and allow these urban centres to access markets less dependent on local demand (Philippe, Léo and Boulianne 1998).

This can be verified for the case of high-technology and knowledge-intensive jobs in Luxembourg using the generalisation made by Zipf (1949). According to Zipf, the relationship between the demographic size of a city (Pi) and its rank in the urban hierarchy (Ri) can be expressed as $Pi = k / Ri^q$, where *k* is a parameter close to the population of the largest city, P1, and *q* is an exponent generally close to 1. By adopting logarithmic scales, the relationship between the rank and the size of cities is expressed linearly as follows: log (Pi) = log (k) - q (logRi). Such a system can be solved using ordinary least squares, where *q* represents the degree of inequality of the size of cities within an urban system (slope of the regression line). Where q (non-standardised) is greater than 1, the collection of cities shows primatial organisation, while if q is less than 1, the organisation in question is polycentric. By replacing the population variable (Pi) by a measure of the employment in each of the 18 urban units identified at the level of Luxembourg, the global distribution of jobs by urban unit is in accordance with a rather primatial organisation (slope k = 1.41), which is strongly increased when only those jobs linked to knowledge, as identified in this study, are considered (1.64) in comparison to other jobs (1.33). This fact is marked by a correlation coefficient R^2, which is lower in the case of jobs linked to knowledge relative to total employment and other jobs (Table 1.3).

Table 1.3 – Distribution of jobs by rank of urban units, 2005

	Slope q	Coefficient R^2
Rank – total population	1.32	0.98
Rank – total employment	1.41	0.79
Rank – high technology and knowledge-intensive employment	1.64	0.65
Rank – other employment	1.33	0.86

Source: IGSS. Calculations: Sohn and Walther.

Note: The urban units consist of agglomerations and isolated towns and are defined by Sohn (2006).

These findings confirm that the category of jobs linked to the knowledge economy is distributed in a clearly unequal way in accordance with the rank of national urban units: the primatial agglomeration

(Luxembourg) is distinguished by a disproportionately high number of jobs of this type in relation to the agglomeration with the rank immediately below (Esch-sur-Alzette) and the following ones. This can be used as supplementary validation criteria for the classification used in this study.

Dynamic of jobs associated with the knowledge economy in Luxembourg

The temporal and spatial development of high-technology and knowledge-intensive employment in Luxembourg is analysed from two perspectives. Firstly, a sectoral analysis is made of the development of the number of high-technology and knowledge-intensive jobs from 1994 to 2008 at the national level. Secondly, an analysis of the spatial distribution of people employed in Luxembourg is carried out on the basis of the location of the head office of the enterprise.

Growth of high-technology and knowledge-intensive jobs

Between 1994 and 2008, the average annual growth in high-technology and knowledge-intensive jobs was particularly significant in Luxembourg (+5.5%) and was markedly higher than that in other sectors of activity (+3.1%). Over this period, the number of jobs linked to the knowledge economy rose from 61,675 to 129,697, while other jobs rose from 147,709 to 227,522 (Table 1.4). These figures show that the high-technology and knowledge-intensive jobs represent a growing portion of the total employment within Luxembourg, rising from 29.2% in 1994 to 36.2% in 2008.

Table 1.4 – Persons in employment by category, 1994-2008

Categories	1994	1996	1999	2002	2005	2008	Average annual growth 1994-2008
High-tech and knowledge-intensive jobs	61,675	66,768	79,421	98,621	107,424	129,697	5.5%
Other jobs	147,709	153,779	172,180	193,015	207,867	227,522	3.1%
Unknown	1,685	4,619	1,633	1,049	440	795	/
Total	***211,069***	***225,166***	***253,234***	***292,685***	***315,731***	***358,014***	***3.8%***

Source: IGSS. Calculations: Sohn and Walther.

Note: The "Unknown" category refers to employees without a NACE code.

Analysis of the development of high-technology and knowledge-intensive employment by sector of activity allows the more precise identification of the activities that act as driving forces within the Luxembourg economy (Table 1.5). In absolute terms, these are the financial KIS (40,856), national and international bodies (20,603) and market KIS (28,310), which included the largest number of jobs in 2008. These three sectors represent on their own 69.2% of total employment linked to the knowledge economy. The high-tech KIS (+14.4%) and market KIS (+9.2%) sectors saw the highest levels of average annual growth over the period in question, which is explained by the fact that these sectors accompanied the development of the financial centre.

However, not all the high-technology and knowledge-intensive sectors can be considered as key activities of the Luxembourg economy. In absolute terms, the weakness of the high-tech manufacturing industries, which represent only 0.6% of jobs linked to the knowledge economy, can be mentioned. The rate of growth measured between 1994 and 2008 for other KIS (+4.3%) – including higher education, healthcare and culture – showed weaker growth than the more dynamic sectors of Luxembourg's knowledge economy.

The examination of the sectoral distribution of jobs linked to the knowledge economy and their development gives rise to an uneven economic profile for the Luxembourg metropolis. Its strong points include, unsurprisingly, finance and high added-value services that accompany the development of financial activities. This strong economic specialisation has, however, a reverse side, which is marked by under-representation of employment relating to higher education and culture. These findings provide a reminder that Luxembourg presents a highly atypical sectoral profile in comparison to the EU 15 average. Indeed, the portion represented by financial services (29.4%) and services to enterprises (33.8%) is significantly more important than in Europe, while a deficit can be seen in the areas of medium-tech industry (1.8%) and higher education, culture and healthcare (20.5%), which represent just half of the average proportion across the EU 15 (Brinkley and Lee 2008). These characteristics result from the economic model adopted by Luxembourg over the past few decades, which heavily favoured the development of a highly specialised financial industry (Pieretti, Bourgain and Courtin 2007, Walther and Schulz 2009 and this volume, Chapter 2) and invested only very late in the field of education and scientific research, with the exception of certain manufacturing companies that have provided long-established research centres (Goodyear, Delphi).

Table 1.5 – Jobs linked to the knowledge economy, 1994-2008

High-tech and knowledge-intensive sectors	1994	1996	1999	2002	2005	2008	Average annual growth 1994-2008 (%)
High-tech manufacturing industry	394	387	425	522	683	835	5.5
High-tech KIS	1,930	2,912	4,688	7,310	7,655	12,702	14.4
Market KIS	8,257	9,045	12,548	17,920	20,603	28,310	9.2
Financial KIS	20,633	22,479	26,390	33,415	33,835	40,856	5.0
Other KIS	6,085	6,471	7,302	8,462	9,579	10,987	4.3
National and international bodies	13,503	14,254	16,426	18,406	20,706	20,603	3.1
Independent intellectual workers	3,073	3,420	3,942	4,886	5,563	6,404	5.4
International civil servants[1]	7,800	7,800	7,700	7,700	8,800	9 000	1.0
Total	***61,675***	***66,768***	***79,421***	***98,621***	***107,424***	***129,697***	***5.5***

Source: IGSS except when indicated otherwise. Calculations: Sohn and Walther.

[1] According to STATEC.

Distribution of activities by location of head office of enterprises

Employment linked to the knowledge economy by location of head office of enterprises is very unevenly distributed across the country. As shown in Table 1.6, 82.5% of this employment is located within the Luxembourg Urban Area, 7.6% within the South Region and only 2.0% within Nordstad in 2008. At the municipality level, Luxembourg City alone has almost 85,600 jobs linked to the knowledge economy, while Esch-sur-Alzette is home to just 4,725 (equivalent to 73.2% and 4.0% respectively). It should be noted that the number of high-technology and knowledge-intensive jobs grew significantly in the South Region between 2005 and 2008. This is to be explained primarily by the establishment of back office financial services (+1433 jobs) at Esch-Belval, which increased the financial KIS categories and, to a lesser extent, by the relocation of a public research centre previously located in the capital (+148 jobs).

Table 1.6 – Employment linked to the knowledge economy by urban units, 2008

Urban areas	High-tech manufacturing industry	High-tech KIS	Market KIS	Financial KIS	Other KIS	National and int. bodies	Total	(%)
Luxembourg Urban Area	270	8,371	22,447	37,955	6,873	20,585	96,501	82.5
Luxembourg City	*95*	*6,149*	*19,042*	*33,297*	*6,432*	*20,584*	*85,599*	*73.2*
South Region	30	1,885	2,493	2,135	2,373	12	8,928	7.6
Esch-sur-Alzette	*1*	*433*	*933*	*1,482*	*1,864*	*12*	*4,725*	*4.0*
Nordstad	5	730	268	58	1,264	0	2,325	2.0
Other urban areas	531	1,764	4,395	795	1,725	6	9,216	7.9

Source: IGSS. Calculations: Sohn and Walther.

However, the number of jobs linked to market KIS and other KIS fell over the period in question in the South Region, which shows that these relocations hardly managed to create an environment that promotes endogenous development. This observation hardly supports the policy of concentrated decentralisation followed by the state of Luxembourg whose objective is to develop knowledge-intensive activities outside of the Luxembourg urban area by means of the Plan Directeur de l'Aménagement du territoire (National Development Plan) and the Integratives Verkehrs- und Landesentwicklungskonzept für Luxemburg (Integrated Concept for Traffic and Regional Development). In the country's other urban units, the level of employment linked to the knowledge economy is essentially linked to the high-technology KIS (+800 jobs), which results from the establishment of IT service companies in the peripheral zones of activity well served by motorways.

The polarisation of the metropolitan economy within the capital is linked above all to the very heavy concentration of national and international bodies (99.9% of all such jobs are located in Luxembourg City), financial KIS (81.3%) and, to a lesser extent, of market KIS (64.3%), three sectors of activity which represent the majority of employment linked to the knowledge economy within Luxembourg's urban units. The dominance of Luxembourg City within the metropolitan economy of the country has traditionally been reflected by a higher proportion of employment linked to the knowledge economy within the total number of jobs (48.1%) than elsewhere in the country (Map 1.1). Leaving to one side the capital, the municipalities that display a relatively high rate of knowledge-intensive employment are principally located within the Luxembourg agglomeration (in particular Niederanven, Sandweiler,

Hesperange, Bertrange and Strassen), with the notable exception of Ettelbruck, which owes its score to the presence of a major hospital facility. In the South Region, 16.5% of jobs are linked to the knowledge economy, with higher figures for the areas of Esch-sur-Alzette (26.9%) and Bascharage (22.9%).

At the national level, the disequilibrium between the Luxembourg urban area on the one side and the South Region and Nordstad on the other is manifested both in the total number of jobs and with regard to the capacity for innovation and thus wealth creation that this represents (Figure 1.3). The development of a metropolitan economy based on the financial KIS, the national and international bodies and the market KIS is responsible for this development. These findings confirm the analyses conducted by Walther and Dautel (2010), which showed that these activities are currently undergoing a certain level of decentralisation in favour of the Luxembourg agglomeration, linked to the expansion of the city centre and the emergence of a metropolitan area strongly structured in functional terms by the development of high-order services within the Luxembourg urban area.

Map 1.1 – Employment by location of head office of enterprises and proportion of high-technology and knowledge-intensive employment, 2008

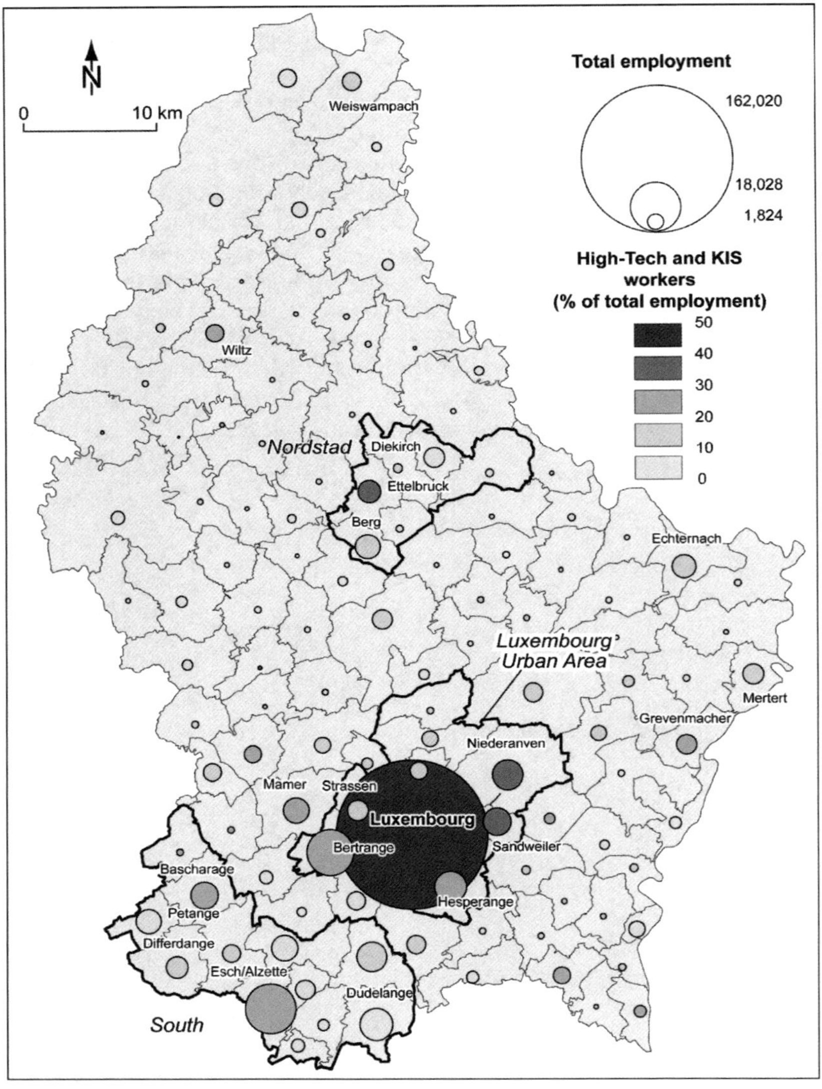

Sources: IGSS. Cartography: Walther.

Figure 1.2 – Proportion of jobs linked to the knowledge economy by location of head office of enterprises, 1994-2008

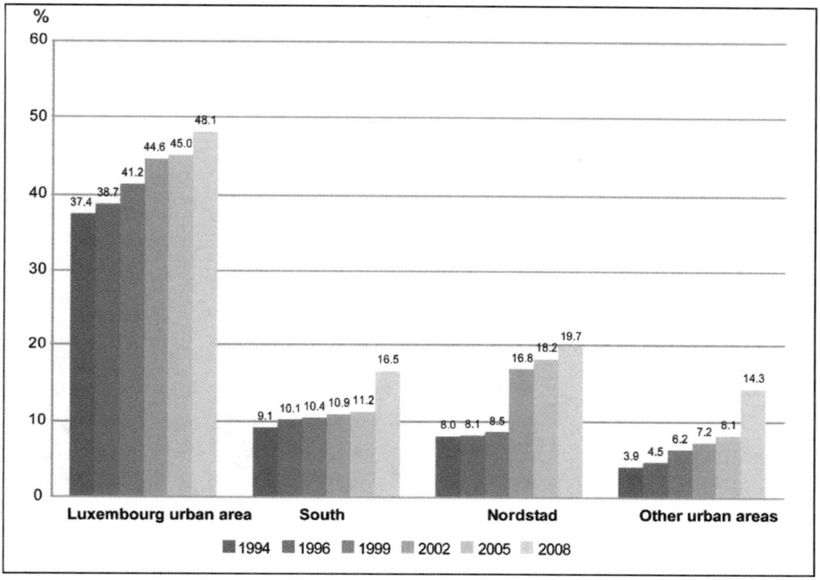

Sources: IGSS. Calculations: Sohn and Walther.

Note: the variations in the rate of metropolitan employment for Nordstad are explained by the changes in classification from 2002 onwards.

Distribution of jobs linked to the knowledge economy in Luxembourg City

The preceding aspects have demonstrated that Luxembourg City is by far the country's leading location for employment linked to the knowledge economy and shows a significantly higher proportion of these jobs in relation to other urban units. An analysis of the jobs made on the basis of real place of work of working persons in 2005 confirms this observation and shows that the jobs linked to the knowledge economy are also unevenly distributed in terms of category at the intra-urban level (Table 1.7).

Table 1.7 – Distribution of jobs by real place of work in Luxembourg City, 2005

Sectors	**Jobs**	
	Number	Proportion (%)
Knowledge economy	56,986	46.2
High-tech manufacturing industry	*88*	*0.1*
High-tech KIS	*2,959*	*2.4*
Market KIS	*12,207*	*9.9*
Financial KIS	*28,444*	*23.1*
Other KIS	*1,507*	*1.2*
Extra-territorial organisations and bodies	*11,781*	*9.6*
Other jobs	66,275	53.8
Total	**123,261**	**100.0**

Source: STATEC. Calculations: Sohn (2009b).

As shown by Map 1.2, the financial services and services to enterprises, as well as extra-territorial organisations, are also heavily concentrated in certain areas, while high-technology services and other services linked to higher education, culture, sport and healthcare are distributed in a much more homogeneous way (the high-technology manufacturing industries representing a negligible proportion of employment). This characteristic can be explained by the heavy specialisation of Luxembourg City in finance, and by its status as a European capital. These characteristics have led (during the history of the development of the capital) to phenomena of agglomeration typical of a city for finance and typical of administrative districts for the European institutions (Gerber and Klein 2009).

Map 1.2 – Location of jobs linked to the knowledge economy in Luxembourg City, 2005 (see colour section)

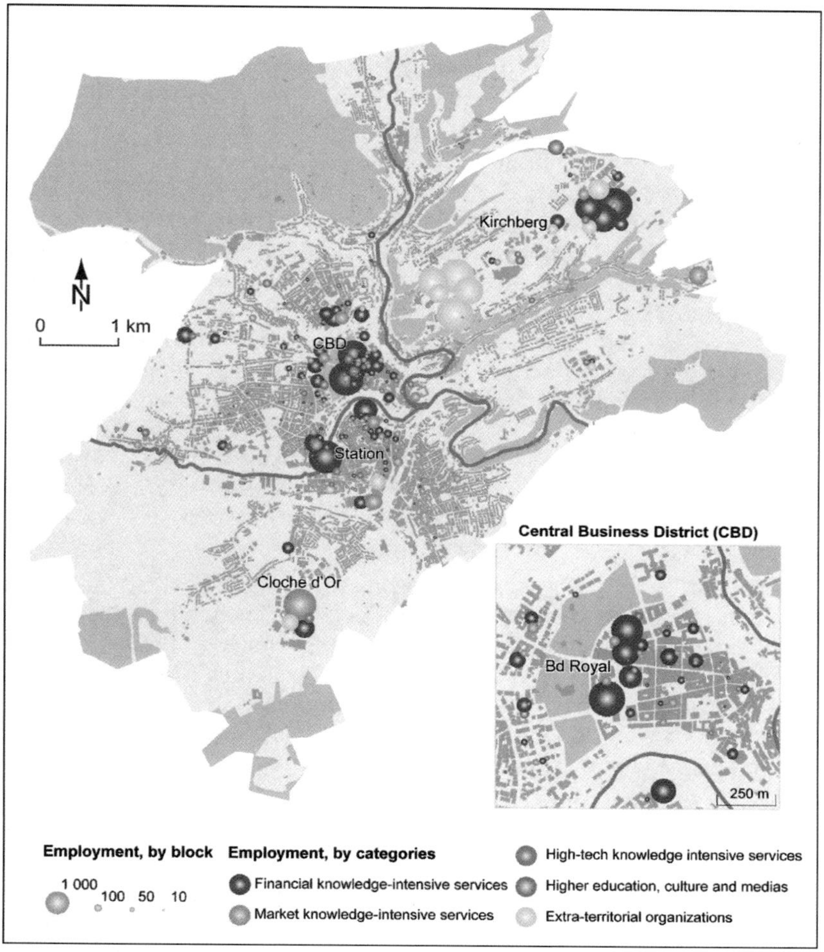

Source: STATEC

Cartography: Sohn (2009b), modified by Walther.

Financial activities represent 23.1% of the city's employment. Their spatial distribution shows two principal poles: the Ville Haute district, home to the Central Business District (CBD), and the Grünewald district in the far northeast of the Kirchberg Plateau. The CBD houses 6,800 jobs linked to financial services, as well as 57 banks out of a total of 156 in 2006. It is no longer limited to the historic Boulevard Royal, but rather extends across the whole of the old town and to a large extent to the west of the former fortifications of the latter and across the north part of the Bourbon Plateau. Apart from the prestige associated with the historic location, the density of establishments and their proximity to the organs of state, to the Luxembourg Stock Exchange and to the sectoral organisations (Luxembourg Bankers' Association, Luxembourg for Finance, Association of the Luxembourg Fund Industry) facilitates face-to-face contact.

As the city centre is small, it is commonly observed that firms wanting to house all of their activities together within a large building such as Commerzbank (2004) and UBS (2008) choose to locate themselves in Kirchberg. Kirchberg offers conditions favourable to the development of financial activities, in particular for the largest banking groups (Clearstream, BGL, HypoVereinsbank, Deutsche Bank, Deka Bank, Commerzbank) and insurance groups (Foyer) that are seeking office space suitable for their activities. This part of the city, which was home to 6,600 jobs linked to finance in 2005, is also in the immediate vicinity of the country's road and air infrastructures, which facilitates the movement of employees across borders and business travel. Outside of the CBD and Kirchberg, certain banks have chosen more peripheral locations, around the Limpertsberg Glacis (west) or along the Esch road and the Cloche d'Or business park (south).

Services to enterprises, which represent almost 10% of the city's employment, show different choices of location depending on the size of the companies concerned. While the major accountancy firms are located at central sites (KPMG) as well as peripheral sites (PWC, Deloitte), the smallest companies, with less than five employees, tend to choose residential areas around the centre (Belair, Merl, Bonnevoie, Limpertsberg) with their luxury buildings.

The location of employment linked to extraterritorial organisations reflects the massive presence of European institutions in Luxembourg City, which employ on their own 10,870 people out of the 11,781 within the sector. These institutions are principally located in the west of the Kirchberg district, whose notable inhabitants include the Secretary-General of the European Parliament, certain departments of the European Commission, the Court of Justice and the Court of First Instance, the European Court of Auditors, the European Investment Bank and Fund, Eurostat and the EU's translation centre. This part of Kirchberg, conceived in the 1960s along modernist lines, is currently undergoing redevelopment and houses several buildings designed by world-famous architects, such as Ricardo Bofill (Place de l'Europe), Christian de Portzamparc (Philharmonie), Ieoh Ming Pei (Mudam) and Dominique Perrault (Court of Justice). The development trends governing the establishment of the European quarter were linked above all to political decisions. Thus, the bringing together of the principal European institutions was decided over the course of the urban development planning of the Kirchberg plateau. There were two reasons for this: it permitted the bodies and services that work together to be located together and it created a strong symbol to reinforce Luxembourg's European mission. The other services associated with the extraterritorial organisations are secondarily located to the east of Kirchberg, in the area occupied by the major financial institutions (European Commission, European Investment Fund), in the south of the city (European Commission) and in the station quarter (EC Publications Office).

A very small proportion of jobs are associated with services orientated towards high technology (2.4%) and services connected with higher education, culture, sport and healthcare (1.2%), which are distributed in a homogeneous way across the city. The weakness of this latter category is explained in part by the fact that a certain number of higher education and research bodies are located outside Luxembourg City (Walferdange, Esch-sur-Alzette), and also by the fact that the University of Luxembourg is a very recent creation (2003). In addition, Luxembourg does not constitute an international centre of excellence in the fields of cultural production, sport or healthcare. The jobs in these sectors concern primarily the city's museums and libraries located in the historical centre and in Kirchberg. This relative weakness of sectors that are nonetheless of strategic importance to the city's influence underlines the lacunae of a metropolis whose attractiveness is for now based essentially on activities associated with the financial centre, the true motor of the development and metropolitan economy of Luxembourg.

Conclusion

The functional indicator developed in this chapter on the basis of the work of the OECD-Eurostat suggests that Luxembourg has progressively established itself as a metropolitan region in which knowledge-intensive services are strongly represented and contribute to determining the development path. From the sectoral point of view, the results of these studies demonstrate the importance of the economic transformations that have affected Luxembourg since the mid-1990s. Between 1994 and 2008, the knowledge-intensive jobs have seen stronger sectoral growth than other jobs. The proportion of national employment represented by knowledge-intensive jobs has increased significantly, from 29.2% to 36.2% of all jobs over the period in question. A very marked disequilibrium characterises the different categories of these jobs, with almost seven out of 10 jobs belonging to financial services, national and international institutions and services to enterprises. This situation clearly reflects the heavy specialisation of Luxembourg in financial services and activities that accompany the development of the financial centre, as well as its role as capital of the EU. On the other hand, employment linked to high-technology manufacturing industries compared to activities relating to higher education and culture appears under-represented despite the recent efforts made by central government and Luxembourg City to develop these aspects.

From the geographical point of view, the analysis of employment by location of head office shows that the Luxembourg urban area constitutes the regional motor of the metropolitan area: almost 82% of knowledge-intensive jobs are associated with it. Luxembourg has thus developed, on the basis of certain comparative advantages, in a highly specialised way, which has resulted in an ever-growing number of banks and associated financial services locating there, in turn rendering the financial centre ever more attractive to other organisations within the sector on the global scale. In the rest of the country, the growth of knowledge-intensive employment is linked more to the active and decentralising policies of the central government, as in the cases of Esch-sur-Alzette and Nordstad, than real potential for endogenous development.

The analysis of employment by the real place of work, carried out on the intra-urban area of Luxembourg City, confirms the heavy specialisation of the economy in the financial services and enterprise sectors (representing 33% of the city's total employment), as well as in the extraterritorial institutions, which make up almost 10% of jobs. The numerical significance of these activities contrasts with the other sectors representative of the knowledge economy, such as the high-technology industries and services or the services linked to higher education, culture, sport and healthcare. There is no doubt that this view, dating from 2005, is in slight contrast with the recent efforts made by the City and the State of Luxembourg to develop these sectors. They are doing this through development of major cultural infrastructure (Mudam, Philharmonie) and cultural events of international significance (Luxembourg and Greater Region: European Capital of Culture 2007), as well as by major investment in the University of Luxembourg, which is today undergoing strong growth and is highly orientated towards research.

These specific findings should lead to more research dedicated to knowledge-intensive employment by place of residence. As demonstrated by Sohn and Decoville in this volume, the effects of the polarisation of the workforce are far from limited to national boundaries. Indeed, the strong growth of employment in general, and of employment within the knowledge economy in particular, have rendered the Luxembourg economy ever more dependent on a cross-border workforce that is active in both high-order services and in support for the latter (in the fields of fast food, home delivery, cleaning and security, for example). The transformation of the Luxembourg economy towards a knowledge economy also contributes to reinforcing the integration of the French, German and Belgian peripheral areas, although the methods used in this chapter have not allowed this sectoral development to be measured, as only employment within Luxembourg has been considered. Further research would thus consider an extended sphere, taking into account the regional centre of employment creation, and the peripheral centres located in France, Germany and Belgium.

CHAPTER 2

Financial Services Clustering in Luxembourg City

Olivier WALTHER and Christian SCHULZ

Recent decades have been synonymous with profound changes in the geography of financial centres. Encouraged by major mergers and acquisitions (M&A), the development of communication technologies, deregulation of national systems and globalisation of finance has led to increased competition between international financial centres (IFCs), (Cassis 2008, Faulconbridge 2004, Klagge 2009), and a higher concentration of financial services in a small number of global centres (Grote 2007). Meanwhile, the process of the deregulation of international finance that has occurred since the 1980s has led to the emergence of a number of sub-centres where financial activity is generally favoured by transparent regulatory bodies, high quality infrastructure communication and low taxation. These financial centres, often known as off-shore financial centres, have benefited from the high mobility of capital. Unlike bigger financial centres that can enjoy large economies of scale, these financial markets have gradually specialised and developed specific skills. This is the case in Geneva, for example, whose banking activity relies on private banking and trade finance; in Zurich and Hamilton (Bermuda) with regard to insurance and reinsurance; in Doha (Qatar) for infrastructure finance; in Manama (Bahrain) for Islamic finance; in George Town (Cayman Islands) for international banking and mutual funds, or in Luxembourg for investment funds (Kruse 2005, Roberts 1994, Walther and Schulz 2009).

The sustainability of these financial centres is now challenged by the financial crisis and increasing pressure from the Organisation for Economic Co-operation and Development (OECD), with questions raised about their existence in territorial niches. Thus, some financial centres strongly oriented towards the securitisation of financial assets (such as Jersey) or hedge funds (such as the Cayman Islands) are heavily exposed. The pressure exerted by the OECD and its "grey list" of countries that are not yet compliant with its tax cooperation rules comes on top of this and leads to "tax havens" rushing to sign different agreements with other OECD countries. From a structural point of view, these financial centres are forced to innovate constantly to maintain sufficient autonomy and develop their highly specialised activities.

Given the importance of such sub-centres both for the global financial system and the national economies where they are located, it is surprising that their evolution and specificities have attracted little attention from geographers. Luxembourg, whose development is the subject of this article, is no exception. Some studies conducted at a European level have continued to ignore the city (Karreman, 2009, Tschoegl 2000), whereas other worldwide studies do not comment on the specificities of the location (Poon 2003, Poon, Eldredge and Yeung 2004). As Hübsch (2004), Franz (2005), Pieretti, Bourgain and Courtin (2007) and König *et al.* (2007) recalled, however, Luxembourg presents certain distinctive features that deserve more attention, in particular its flexibility and responsiveness, which has allowed the city to become a hub for the distribution of investment funds at the global level.

This article attempts to fill this gap. The main objective is to question the sustainability of the financial structure established in Luxembourg. In a context where sub-centres are struggling to redesign their focus, we examine how the morphology of the Luxembourg IFC contributes to the sustained comparative advantage of the financial centre and detail the main advantages and disadvantages of Luxembourg as a business location for the finance industry. In doing so, we wish to point out the fact that, far from being limited to a niche policy, the development of the sector is also increasingly dependent on external economies linked to the metropolitan area. Our hypothesis is that sub-centres such as Luxembourg are part of a new environment where the difference will be less and less about regulatory issues and more and more about the diversity and quality of services offered by local actors. This issue is addressed by questioning the two major clichés attached to the financial centre of Luxembourg.

Luxembourg is often considered a fragile IFC because the finance industry, which accounts for a third of the national wealth, is based on certain typical advantages of a niche policy, notably banking secrecy but also several tax laws (see Pieretti, Bourgain and Courtin 2007 for a survey of the legal and regulatory framework). These advantages could be threatened or annihilated by the evolution of the international financial system. Indeed, as with all small states, Luxembourg is strongly dependent on the mobility of capital at the global level (Armstrong and Read 1998) and, as with all off-shore markets, on evolutions in the regulation of on-shore financial markets. Thus, it has to incessantly innovate in such a way that the deregulations that have taken place since the 1980s do not erode its comparative advantages. Until now, the strategy of continuous innovation in Luxembourg has seemed successful, as witnessed by the adaptive efforts made in the area of investment funds, private banking and financial engineering for the past 20 years or so, as well as more recently in the fields of estate management, securitisation, venture capital and reinsurance (OECD 2008).

Luxembourg is also considered a financial centre without real autonomy. The massive presence of foreign banks in the city has sometimes led commentators to judge Luxembourg as a satellite of neighbouring countries, in particular Germany. It is often assumed that Luxembourg is too dependent on the domestic markets of its neighbours to develop as an autonomous financial market. As the financial activity and diversity of concerned establishments grow, some results, however, suggest that the advantages presented by this niche policy cannot by themselves explain the prosperity of Luxembourg. Luxembourg is still often thought of as a tax paradise. However, econometric studies have shown the importance of agglomeration effects and seem to confirm the presence of a "financial cluster" also fed by endogenous forces and centred on banking activities (Pieretti and Bourgain 2003, Bourgain and Pieretti 2006). The growth of the financial sector has produced a significant increase in business services and non-financial market services (downstream industries) and a significant effect of IT services and business services can be observed on financial services (upstream industries). Other studies have suggested that the Luxembourg-based financial industry has led to the development of highly specialised regional cross-border labour pooling (Walther and Dautel 2010). A similar evolution was noted by Hampton (1996) in his analysis of Jersey. Even though the initial regulation was the dominant factor for the bank's move to Jersey in the 1960s, the rapid growth of banking was accompanied by an expansion of associated services, which consecutively expanded the financial infrastructure of the island.

The remainder of this paper proceeds as follows: in the next section, we expose the conceptual framework of our work, including a critical discussion of the cluster concept applied to financial centres; in the third section, we briefly describe the data available in our survey and the methodology; the fourth section analyses the results of our survey, building on interviews and a questionnaire inspired by previous studies conducted in London by Taylor *et al.* (2003); and in the last section, we conclude with a summary of our key findings.

Conceptual framework

Following Gordon and McCann (2000) and Schamp (2009), three main types of factors influencing the development of IFCs can be identified: the nature of financial products and their importance in the global value chain; classical urban externalities; and social networks.

First, it is obvious that the type of financial product developed and traded in a financial centre exerts a considerable influence on its spatial configuration (Faulconbridge *et al.* 2007). As shown by the product information-based approach in finance geography, "financial products often have a distinct spatial configuration of information embedded in their design" (Clark and O'Connor 1997: 95). Three types of products can be identified: transparent products such as change currency that tend to be extremely concentrated in a limited number of large IFCs to benefit from scale economies; opaque products that require much more non-codified knowledge, such as M&A, and that can also be designed in smaller financial centres where local information is crucial; and translucent products such as hedge funds, which are somewhere in between. Considering that opaque products have a tendency to be transformed into more transparent products by rating agencies, regulatory and exchange organisations, financial information brokers or research institutes, second tier financial centres can maintain their competitive advantages in two ways: by developing their intermediation capacity as gatekeepers who mediate between outsiders from global markets and local knowledge; and by transforming transparent products into translucent products (Faulconbridge *et al.* 2007).

Second, as with any other agglomeration of economic activities, financial centres are influenced by classical Marshallian externalities. In the context of a concentration of financial activity at a global level, labour market pooling, the supply of intermediate goods and technological and informational spillovers play a fundamental role (Martin 1999, Porteous 1999). Those effects are particularly obvious in certain financial services, where skills are acquired through training and where face-to-face communication with clients is determinant (Storper and Venables 2004). Agglomeration economies favour those markets where the density of employment and enterprises is the greatest. However, the growth of certain smaller financial centres can also result from the extreme specialisation of certain services, promoted by the exploitation of a niche of national sovereignty. Certain comparative advantages can then lead to a cumulative process in which the installation of banks that are attracted by framework conditions can subsequently increase the attractiveness of the location for banks that are not yet installed there. Arthur (1994), Grote (2008), Grote, Lo and Harrschar-Ehrnborg (2002), Lo (2003) and Clark and Wójcik (2005) all showed that in a situation of increasing returns, those financial centres that possessed a historical advantage kept their pre-eminence. This effect generally results in the creation of a financial centre whose heart is made up of the banks and around which advanced services enterprises come and graft themselves, directly or indirectly linked to finance, among which are found firms active in the fields of audit, law, consultancy and data processing.

Third, IFCs are not just characterised by a concentration of firms. They also connect individuals who share social and cultural values and interact within similar social networks (Thrift 1994). The exchange of these values is essential to maintaining trust and reputation, the two pillars on which much financial intermediation activity relies. Social relationships, informal rules and interpersonal networks established in the business are constitutive of markets (Peck 2005). The effectiveness of these standards and rules is greatly enhanced by the physical proximity between the actors, whether in the City of London (McDowell 1997), Manhattan (Longcore and Rees 1996) or the "Westend" in Frankfurt (Schamp 2009).

The concentration of financial services has forced economic geographers to go beyond the notion of localised production systems in the sense of manufacturing and made them increasingly turn towards the services sector. Inspired by the debates around global or world city approaches, particular attention has been paid to knowledge-intensive business services. Here, the study of "financial clusters" has played a major role. Indeed, the application of the term "cluster" to a geographically localised grouping of advanced services producers such as financial activities is problematic, first because of the "lack of clear boundaries, both industrial and geographical" of the cluster (Martin and Sunley 2003: 10) and second because of the obvious differences between manufacturing and services industries. As Schamp (2009) emphasised, these differences are linked to the heterogeneity of financial services compared with industries, meaning that only knowledge-based services need spatial proximity. Financial services are not "delivered" to clients in a vertical way as manufacturing products are, but they result from the collaboration with other services firms in a knowledge-based environment. Accordingly, financial services are much more based on horizontal collaboration between firms because of both the still strong vertical integration of banks and the nature of the innovation process, which often relies on a network of horizontal linkages.

The literature devoted to clusters makes a distinction between several large types. *Marshallian industrial districts* and their Italian variant are more characteristic of small industrial enterprises whose local owners exchange know-how, workforce and information (Piore and Sabel 1984). To that well-known type of cluster Markusen (1996) added: the *hub-and-spoke districts*, dominated by the presence of one or more very large firms that are vertically integrated and linked to a dense network of local suppliers; the *state-anchored industrial districts*, dominated by the presence of a government institution; and the *satellite industrial platforms*, marked by the predominance of legal and fiscal incentives. Markusen's last type is characterised by a small number of large enterprises, the strong dependence of the firms on their parent companies located abroad, strong state investment as well as labour mobility, a rate of enterprise creation and low agglomeration economies (Pandit and Cook 2003). As pointed out by Markusen (1996), empirical studies carried out in the United States, Japan and Brazil have shown that most of the clusters stem from a hybridisation between the types previously defined and are susceptible to transformation over time.

Apart from the abovementioned agglomeration economies and dise-conomies, and the structural varieties of clusters as classified by Markusen, cluster research in economic geography has underlined the importance of a cluster's dimensions. This allows for a more dynamic and detailed analysis of the locational advantages of a spatially clustered industry. According to Malmberg and Maskell (2002), the aforementioned *horizontal* and *vertical* dimensions of a cluster are complemented by the following:

– An *institutional* dimension: the cluster's embeddedness in a setting of specific institutions, not only comprising intermediary organisations (e.g. business associations, trade unions and development agencies) and other formal institutions (e.g. laws, codified regulations and norms), but also informal institutions (see North 1990, Hodgson 2006) such as the "rules of the game" and conventions specific to the industry.

– An *external* dimension as a crucial precondition for the cluster's long-term success. Only those clusters embedded in interregional/ international networks will frequently receive innovative impulses from abroad via cooperative projects, trade fairs, conferences and personal relationships. Bathelt, Malmberg and Maskell (2004) introduced the notion of "global pipelines" as being the external communication channel for local clusters, influencing the so-called "local buzz" typical for clusters.

In his paper questioning the cluster concept applied to financial centres, Schamp (2009) enlarged this perspective using a multiscalar and multidimensional approach. Taking into consideration the case of Frankfurt, he argued that the value chain mostly depends on the national level and its legislative setting (Germany) and urban economies become relevant at the scale of the agglomeration (Frankfurt/Rhein-Main), whereas the importance of social networks is clearly bound to the financial district scale (Westend). Even though this paper empirically focuses on the two latter scales, namely the agglomeration of Luxembourg and its intra-urban neighbourhoods, where close to 95% of enterprises and employment were concentrated in 2005, the national and superior scales have to be included in our analysis.

Besides the national scale and the Grand Duchy's particular regulatory framework, the international scale seems to be of particular importance, at least in two aspects. First, the convergence of regulations within the

European Union, fostered by the current crisis-led debates on a reregulation of the financial markets, is supposed to be of tremendous importance for the future development of all niche economies in this sector. Second, the strong dependency and narrow functional integration of most foreign banks in Luxembourg with their mother companies especially in Germany (Frankfurt) and the UK (London) has to be considered part of the central value chain when talking about its localisation or territorial boundness.

Regarding the more local dimensions of the financial centre, the following research questions guided our study:

- What are the main trajectories of Luxembourg's emergence as a financial centre?
- What is the importance of industry-specific agglomeration effects and spatial proximity in this "cluster"? How strong are the local horizontal and vertical ties?
- How can the local spatial organisation be explained and what potential changes should be expected?
- To what extent has Luxembourg – as a presumably offshore financial centre – developed endogenous competitive advantages going beyond regulatory incentives and amenities?

Furthermore, we will try to discuss the aforementioned changes at the international level and Luxembourg's capacities and potentials to adapt to these turbulent environments. This leads to the overarching question: to what extent does the financial centre have a sustainable perspective?

Data and methodology

In 2005 the financial sector in Luxembourg comprised 424 firms: 128 banks, 110 insurance and reinsurance companies, and 185 professionals of the financial sector (PFSs), i.e. enterprises either connected or complementary to the financial sector. Drawing inspiration from the study carried out by Taylor *et al.* (2003) on the City of London, a questionnaire allowing for the identification of the concentration and dispersion factor was sent to these 424 enterprises. From these questionnaires sent, 109 were returned (25.7%). The sample thereby consists of 44 banks, 22 insurance and reinsurance companies and 43 PFSs.

The study was completed with 22 personal interviews with top managers. These came from a sample of the 100 global service firms identified by Taylor (2004) in his Globalisation and World Cities study. In 2005, 48 out of these 100 global service firms were located in Luxembourg. Interviews were conducted with banking and finance firms (10) as well as insurance (1), accountancy (6), law (1) and management consultancy firms (4).

Based on Krugman (1998) and Pandit, Cook and Swann (2001), the notion of concentration and dispersion factors is used when trying to gain insights into the locational preferences of the interviewed firms. As far as concentration forces are concerned, we distinguished between market size effects (i.e. backward/forward linkages), the benefit of a big labour supply, pure external economies such as technological and knowledge spillovers and regulation factors. Dispersion forces were identified as land rents, labour costs, pure external diseconomies, such as transports costs, congestion and density, and regulation framework (Table 2.1).

Table 2.1 – Concentration and dispersion forces

Concentration	**Dispersion**
Market size effects	**Land rents**
Ability to rely on a large array of supporting services	Cost of business premises
Ability to find firms who will supply customised services	Rental charges (flats and houses)
Proximity to market-leading customers	
Customers find it easier to locate/interact with you	
Credibility of the address for your company	
Access to the European Union market	
Proximity to leading competitors	
Best place to be to take market share from rivals	
Labour market	**Labour market**
Access to a skilled labour supply	Labour costs
World class teaching and research centres	
Access to educational, cultural or sports facilities for employees	
Access to international schools	
Availability of staff with language skills	
Pure external economies	**Pure external diseconomies**
Ability to access real time information on market trends	National/international transport links
Proximity to a relevant exchange or physical marketplace	Price of telecommunications
Knowledge transfer from the City's financial atmosphere	Environmental quality and pollution
	Local infrastructure (including buildings)
Regulation framework	**Regulation framework**
European Directives are rapidly put into national legislation	Government regulation
Simple and favourable legal and regulatory environment	Administrative burden
Authorities can be approached easily	
Support from service sector institutions	
Support from city government	
Proximity to professional bodies	

Sources: Krugman 1998, Pandit, Cook and Swann 2001.

Before the results of the questionnaire and the qualitative interviews are presented and discussed in more detail, the historic evolution of the financial centre and its current structure will be summarised.

Financial centre of Luxembourg

According to the Global Financial Centres Index, computed from 47 quantitative indicators and surveys (City of London 2008), the position occupied by Luxembourg among the 50 largest financial centres in the world is strengthening, which is a trend common to several smaller niche centres such as Geneva, Jersey and Guernsey. The city stood in 15th place in September 2008, making it the seventh largest financial market in Europe after London, Zurich, Geneva, Frankfurt, Dublin and Jersey. It is considered the second world financial centre with regards to assets managed by undertakings for collective investment (UCI) after New York, the first captive reinsurance market in the EU and the first private banking centre in the Eurozone (Luxembourg for Finance 2009). These rankings, however, do not take into account that competition among financial centres generally happens vertically: between financial centres of different levels. As Grote (2009) showed, most of the competition in Europe takes place between London and second-tier financial centres such as Paris, Frankfurt, Amsterdam and Milan on the one hand, and between these second-tier centres and national or offshore sub-centres such as Munich, Hamburg, Zurich, Geneva and Luxembourg on the other.

Contrary to other financial centres, whose development has been based on external scale economies, the strong concentration of financial enterprises witnessed in Luxembourg is first explained by the legal and tax incentives that are characteristic of a niche policy. Since the 1960s the main activity of foreign banks has been linked to international credits and marked by the development of Eurodollars and holding companies. Although London specialised in issuing Eurodollars in favour of international banks from the early 1960s, Luxembourg developed into an international centre for the administration of these currencies (OECD 2008). The impressive settlement of foreign banks has been made possible by a technological development that allows banks and financial institutions to become global and develop new financial products (Martin 1999). These changes explain why the first foreign banks established in Luxembourg were primarily concerned with meeting the demands of customers wanting to borrow foreign currencies such as pounds and dollars while escaping the banking regulations imposed by states. American banks (Citibank, J.P. Morgan) were the first concerned, followed as early as the middle 1960s by German banks willing to avoid statutory reserves (Dresdner Bank, Commerzbank, Deutsche Bank, Deka Bank) and Swiss banks

(UBS, Credit Suisse) attracted by the growth of the Euromarket and the presence of German banks (Kirsch 1994).

The introduction of floating exchange rates in the early 1970s caused a decrease in the controls of international capital flows and a potential weakening of Eurocurrencies. However, the Eurodollar market has since then been supplied by demand from oil-producing countries willing to invest their reserves in European banks (in Eurodollars) rather than in US banks, fearing a possible freezing of their assets in the United States. Some US banks have selected small tax havens in the Caribbean to minimise their costs, whereas banks from Germany (DZ Bank), Italy (Banca Nazionale del Lavoro, Capitalia) and Scandinavia (Swedbank, Nordea Bank, Danske Bank, Svenska Handelsbanken) have chosen Luxembourg. The international credit crisis during the second half of the 1970s, however, hastened a progressive reorientation of the banking market towards private banking, asset management and the investment funds industry. The beginning of the 1990s was marked by the second massive arrival of German banks (WGZ, Sal. Oppenheim, Landesbank Berlin, Bayerische Landesbank) attracted by the swiftness with which European laws are transposed into domestic law, as well as by the inventiveness of establishments in offering adapted financial products. At that time, Luxembourg hosted some 220 foreign banks, more than Paris or Frankfurt (Laulajainen 1998). As Figure 2.1 indicates, the number of banks was highest in 1994 (222) before undergoing a constant decrease because of M&A that marked the sector at the international level (Grote 2008). Among the 156 banks that made up the sector in 2007, 26% were foreign branches and 69% subsidiaries. The diversity of origins increased with the arrival of Italian (Banca Antonveneta, Banca Lombarda, Fideuram), Brazilian (Unibanco), Chinese (Bank of China), Israeli (Hapoalim, Leumi) and Portuguese banks (Santander, Itau).

Figure 2.1 – Evolution of the number of banks and of their balance sheets, 1960-2008

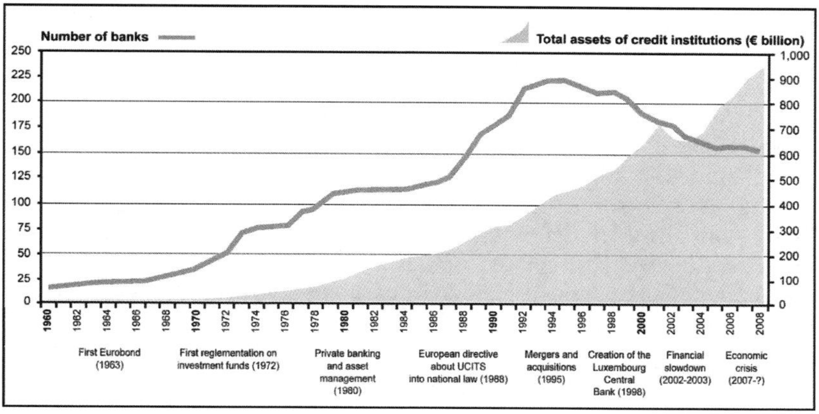

Sources: Central Bank of Luxembourg; CSSF. Illustration by Walther and Schulz.

From 2005 the possibility for e-commerce companies to process the electronic payments of their customers through Luxembourg, thanks to an electronic bank licence, helped diversify the activities of the financial centre. Amazon, eBay, Skype, PayPal and Digital River have chosen Luxembourg as a central hub for their activities in Europe, but without significant impact on employment in the Grand Duchy. In 2007 family estate management companies were created and replaced the holding companies threatened by the European revision of the law of 1929. During that period the success of the Luxembourg financial market was characterised by the sustained growth of the balance sheets of credit institutions. Certainly, the economic crisis of 2002 and 2003 saw the temporary fallback of balance sheets, but it increased again in 2004, reaching a maximum of EUR 1,002 billion in October 2008.

The consequences of the recent crisis are harder to evaluate. The numbers so far present a mixed picture. It is true that the balance sheets of banks decreased to EUR 848 billion in June 2009 but the financial sector has proven resilient so far. First, few subsidiaries (mostly from Iceland) have closed, with two major banks, Dexia and Fortis, rescued by the government. Second, employment in pure credit institutions decreased by only 1.5% between September 2008 and June 2009 (399 jobs). Third, credit institutions registered a positive variation in their profits before provisions, which increased from EUR 3,050 to 3,396 million between June 2008 and June 2009 (+11.3%), according to the Financial Sector Supervisory Authority. Fourth, the UCI vehicle, the major component of the fund industry that allows collective investment schemes to operate throughout the EU, seems to have retained the confidence of investors: the number of UCIs increased steadily to 3425 in May 2009 but the volume of net assets declined by 18.9% between May 2008 and May 2009.

In 2005 the financial sector (banks, insurance companies, UCIs and PFSs) directly or indirectly represented 19% of domestic employment, 31% of tax returns and 38% of the total domestic added value (Deloitte 2006). Services activities directly connected to financial activity – such as fiduciaries, public notaries and attorneys at law, audit and advisory companies, and companies in charge of the administration of investment funds – are also included within PFSs. As Table 2.2 shows, employment in the sector grew strongly (+5.3%) between 2000 and 2007, particularly for PFSs and UCIs, which benefit from the outsourcing of certain activities previously carried out by banks.

Table 2.2 – Evolution of employment in the financial sector, 2000-2007

	2000	2001	2002	2003	2004	2005	2006	2007
Banks, Stock Exchange, Cetrel	23,321	23,952	23,600	22,839	22,869	23,550	25,084	26,457
Insurance companies	9,268	9,532	9,556	9,579	9,555	9,662	9,769	10,886
PFSs	3,499	4,176	4,399	4,455	6,059	6,547	9,928	12,174
UCIs	0	0	0	98	507	1,572	2,069	2,348
Total	36,088	37,660	37,555	36,971	38,990	41,331	46,850	51,865

Note: Cetrel is a provider of solutions for electronic payments.

Source: CSSF; CAA; Cetrel and IGSS.

The evolution of employment in the banking sector by nationality was strongly influenced by the increase of cross-border and foreign workers (Figure 2.2).

Figure 2.2 – Employment in the financial sector by nationalities, 1960-2007

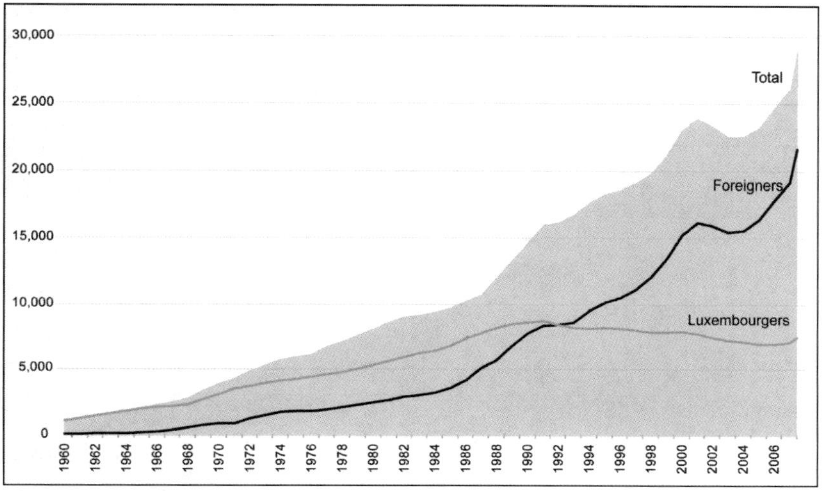

Sources: Central Bank of Luxembourg; CSSF. Illustration by Walther and Schulz.

Although Luxembourgers still provided 90% of the workforce until the late 1960s, their proportion had decreased to 25% 40 years later. The growth in the employment of foreigners has been particularly strong since 1985, when Luxembourg transformed into a private banking and investment funds centre. The growth in the employment of foreigners is subject to greater conjectural changes than that of Luxembourgers. For example, the financial crisis of the early 2000s had a greater impact on foreign workers, whose numbers declined sharply between 2001 and 2003 than on the number of Luxembourgers, which has continued to decline steadily. This might be explained by the fact that Luxembourgers usually occupy positions less prone to changes in international activity (e.g. branches in retail banking) and/or that banks are more reluctant to lay them off in bad times.

The centre of the financial centre, weighted by the number of jobs in each establishment, is located between the Central Business District (CBD), where banks have been historically concentrated, and the Kirchberg area in the east, developed since the 1970s and home to the larger establishments (Map 2.1).

Map 2.1 – Location of the financial establishments in Luxembourg, 2005 (see colour section)

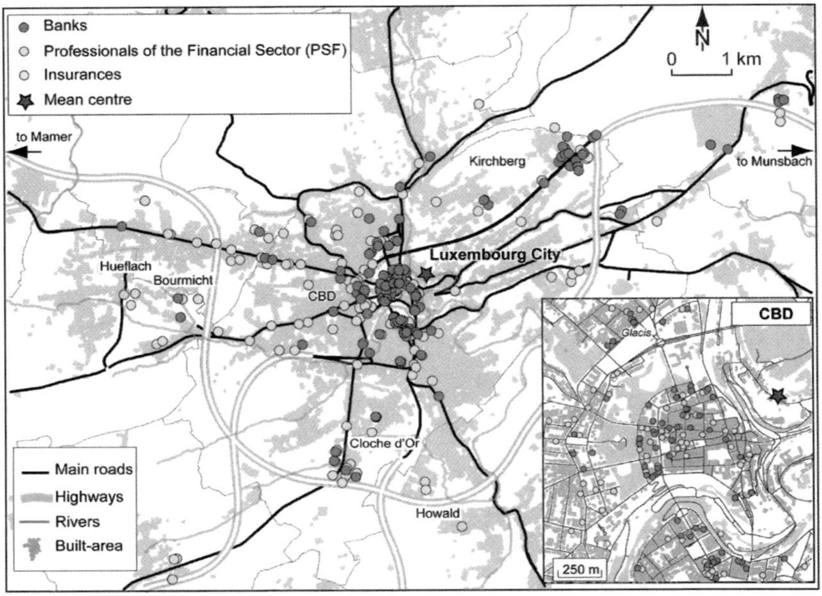

Source: CSSF. Cartography: Walther.

The calculation of the average nearest neighbour index confirms that the financial sector in Luxembourg is made up of establishments that are mainly concentrated in the city and agglomeration of Luxembourg. This index is calculated by comparing the mean distances observed between each establishment and its closest neighbour at those average distances that would be obtained in the case of a hypothetical contingent distribution. The index obtained (0.16) is much lower than 1, which confirms the fact that the establishments of the financial sector, all categories mixed together, have a strong tendency to form localised groups of businesses. The distribution of insurance firms (0.21) presents a higher degree of clustering than that of banks (0.27), whereas that of PFSs is more dispersed (0.38). These results allow for a more detailed analysis of the factors, which in the opinion of enterprises concur either to the concentration or to the dispersion of the sector.

Financial sector clustering

This section presents the results of our survey. Respondents were asked to rank the importance of the different factors from 1 (low importance) to 3 (high importance) with an option of 0 if not relevant.

Legal incentives and spillover effects

With regard to concentration forces, the results clearly show the importance of the legal and regulatory environment as well as the proximity to national authorities, two qualities for which Luxembourg is traditionally known (Figure 2.3).

This aspect is yet again strengthened by the importance given by surveyed enterprises to their proximities to professional institutions and to those of the City of Luxembourg. This contributed to a certain institutional thickness, i.e. "interinstitutional interaction and synergy, collective representation by many bodies, a common industrial purpose and shared cultural norms and values" (Amin and Thrift 1995: 102).

Figure 2.3 – General advantages of a Luxembourg City location

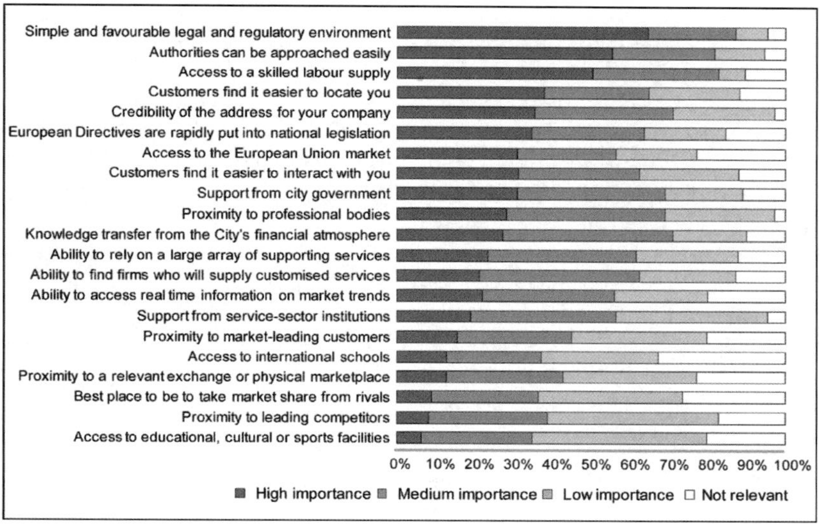

Note: N=109 firms. Source: Walther and Schulz.

Our results also confirm the importance granted to the adaptability of its legislative and regulatory framework. Luxembourg distinguishes itself by a first mover advantage where European directives are rapidly transposed into national law. This allows Luxembourg to become the first country of the European Union to apply the regulation on the Undertakings for Collective Investment in Transferable Securities (UCITS I), encouraging the domiciliation of investment funds as early as 1988. Other regulatory modifications to international pension funds and new financial engineering products intended to replace holding companies dating back to 1929 have also strengthened the sector. In addition to holding companies – which provide tax exemption on profits and appreciations and an absence of withholding at the source – are the financial participation commercial companies (i.e. *SOPARFI*), which benefit from conventions against double taxation. There are also UCIs such as open-ended investment companies, which have found favourable conditions of domiciliation in Luxembourg.

On a par with the growth in the number of banks, tax incentives are also doubled through spillover effects. The time of the exclusive tax niche seems to have disappeared. As the Chief Executive Officer (CEO) of one of the largest German banks noted:

> There are some who still think people were coming here with a suitcase full of money and we were opening accounts for every fool. Worldwide, Luxembourg has got one of the most severe laws against money laundering. This hasn't existed for the past 20 years, but Luxembourg still suffers from this image (June 19, 2007).

The financial centre having reached a certain size, the installation of a great number of foreign banks attracts new banks because of either competition or credibility. This effect strengthens the critical mass of the financial market, which then strengthens the financial sector *stricto sensu* in the area of insurance companies, advanced services enterprises and audit, accounting, law or information technologies. As recalled by the CEO of a large Japanese bank:

> Many banks, accountancy and financial service providers are integrating here and they create this image – a sort of brand of Luxembourg fund. So, foreign institutions such as our company used to come here to operate in this kind of business, because other people came here. You know that when you compete with these kinds of people, you have to operate here (June 6, 2007).

Whereas enterprises are unanimous in acknowledging the importance of having a qualified workforce at their disposal, they also acknowledge that it does not represent an important factor in their localisation in Luxembourg, since it is precisely that very workforce that is lacking. As expressed by a partner of a large auditing firm, the availability of the workforce is certainly not an asset of the country:

> It is not because we have 1,400 people here that we can claim there is a strong labour market in Luxembourg. Those people where imported from Marseille, Greece, Portugal, Dublin and so on. It's the most important issue we have as a firm, or any group has here in Luxembourg, but we cannot solve it simply by [recruiting in the country or] in the region (June 15, 2007).

By contrast, the enterprises put forward the multilingual capabilities of the workforce, which offers a true advantage over other global financial markets. Compared with other financial markets such as Dublin, Luxembourg offers interesting opportunities. English, German and

French are all standard. As recalled by the managing director of a German investment enterprise: "If you place a new fund for the German market, if I were in Ireland, I would have to submit it in English to the authorities and then translate it. I don't need it in Luxembourg. Multilingualism is a real asset" (June 19, 2007).

Two more factors are directly linked to clients: the credibility of the address and the ease afforded the clients when finding the financial establishment. Even though the Grand Duchy of Luxembourg is relatively unknown except for its financial activities, these results confirm that financial enterprises place a preponderant importance on being easily accessible.

Another advantage of a Luxembourg location is that the city is very small. However, as Boschma (2005) argued, spatial proximity is neither a necessary nor a sufficient condition for localised knowledge spillovers to develop. In an economic agglomeration, financial firms require cognitive and social proximity, the former being related to the capacity of actors to share the same knowledge base and expertise and the latter being related to the embeddedness of business relationships and trust in a social context. As a small IFC, Luxembourg provides strong advantages on both kinds of proximity: cognitive proximity is encouraged by the specialisation of the financial industry in a limited number of products or operations, whereas social proximity is encouraged by the dense network of formal and informal relationships between peers and/or competitors in the City and between the City and the political/regulatory sphere.

The surveyed firms deem it important to benefit from a transfer of knowledge from the City, but in a less accented way as for those factors put forward above. As recalled by the manager of an American bank in Luxembourg City, active in the field of fund administration, the localisation of foreign branches is strongly dependent on the nature of their activities:

> We are a back office firm. We don't see many clients and, if so, rather institutional clients for whom it doesn't matter whether we are located on Boulevard Royal [a prestigious address]. Our clients are mainly abroad and thus [...] we are organised to meet our clients at their location. One relationship manager is always in the city or the country where the headquarters of our clients are [...] which is not the case for our back office activities which can be everywhere (June 18, 2007).

As this partner from an auditing firm argued, being physically in contact with clients is absolutely necessary for every activity that is built on trust:

> We live with the human contact, because otherwise we would not sell anything, because we have to build trust, we have to see people. It's very important in our business, we are living from that. Because informal contacts are spreading the news, when something is coming up or you get hints of what's going on in companies, so that's a mine of information (June 15, 2007).

Access to the European Union market is a key differentiating factor for the fund industry. As previously indicated, the fund industry initially developed in Luxembourg because of a first mover advantage. Nowadays, being located in Luxembourg offers an opportunity to benefit from low taxes and being able to administrate investment funds in all European markets. However, Luxembourg does not provide comparable access to the EU market as that of London, where most American firms want to be located when they need a European hub.

Inversely, enterprises consider the factors more directly linked to their employees less important. These include the availability of access to international schools or cultural or sports infrastructure and the spillovers that might result from the proximity of competitors, a stock exchange or real time information on financial markets.

Tiny but costly

Among the uppermost factors that negatively affect economic activity in Luxembourg are the constraints presented by labour and real estate costs, as well as certain pure external diseconomies such as transport and travel costs (Figure 2.4).

The cost of labour is an area where the finance industry has serious competition from the civil service. As previously shown, a specificity of Luxembourg is that the financial community is predominantly composed of foreign residents or commuters. The fact that civil servants must speak Luxembourgish tends to encourage nationals to be employed by the state or municipalities rather than by financial firms, a comfortable situation in terms of salaries and employment security. For that reason the cost of labour is regarded as a highly impeding factor for financial firms, especially compared with neighbouring countries. As this head of business management from a Swiss bank recalled: "Indexation and this kind of stuff will kill the marketplace if it goes on like this. The minimum salary is 1,784 Euros, as far as I know. Even just after finishing school, you get 2,000 Euros a month. Who earns 2,000 Euros a month in Germany at the very beginning?" (June 21, 2007).

Figure 2.4 – General disadvantages of a Luxembourg City location

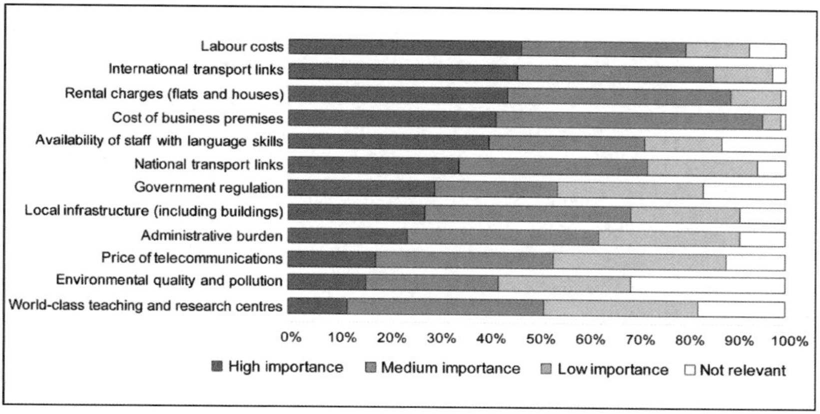

Note: N=109 firms. Source: Walther and Schulz.

At least until the recent financial crisis, finding enough qualified employees was a major task for HR officers in the finance industry. The shortage of employees was so challenging that great efforts were engaged to hire people from all over the world, especially from other large financial centres such as Brussels, Frankfurt, Paris, Madrid, Zurich or Dublin, where many of the Luxembourg-based banks also have offices, as well as from other cities where HR officers had privileged informal links, such as Lyon or Geneva, or from southern Europe countries such as Portugal or Greece.

As many interviewees recalled, attracting a qualified workforce to Luxembourg is not an easy job. Because of the strong competition between financial centres and the poorer international image of Luxembourg as an attractive business location, incentives need to be especially strong (in terms of wages but also of a non-pecuniary kind) for senior managers of the sector to accept coming to Luxembourg. From that point of view, the small size of the city represents a handicap that is only partially compensated by the exceptional level of its infrastructure. As explained by the head of business management at a large Swiss bank: "Who wants to come from Frankfurt to Luxembourg? Nobody! You get [senior bankers] from Luxembourg to New York, to London, even to Switzerland – you don't get them to Luxembourg. Or you pay a lot. That's the price" (June 21, 2007).

Being able to attract senior managers is an important issue in terms of the diversification of the activity and evolution towards fewer back office activities. Core financial activities are organised in a hierarchical and uneven way at the international level. As far as asset management is concerned, for example, the core business is still operated in a few large financial centres such as London, Paris or Zurich for British, French or Swiss banks respectively. Being innovative in this kind of business is certainly an important issue for Luxembourg if it wants to develop more core activities in the value chain of the finance industry (Pieretti, Bourgain and Courtin 2007).

HR officers agreed that there was a lack of experienced people but they also recognised that the situation was not much better in Brussels, Frankfurt or Paris. They also noted that the increasing international recognition of Luxembourg as a specialised financial centre makes recruitment easier than a decade ago, especially for young motivated graduates with up to five years experience. These young bankers are said to find comparative advantages if they agree to work hard. As the senior executive of an audit firm said:

> [These young bankers] want to learn and they are not sure about staying here, obviously, for the rest of their life and that makes it very interesting, because they are better, harder workers. I can see that in our company. Generally, we have very, very good feedback when our people work in other countries for projects. So, we attract young people who are a bit more motivated than the average worker (June 8, 2007).

Office costs also appear to be a central factor. The competition for central space makes Luxembourg one of the most expensive cities in Europe regarding office real estate. The total occupancy cost per workstation was EUR 9,360 in 2007, comparable to cities such as Seoul, Munich or Zurich (DTZ 2007). The *ex nihilo* creation of the Kirchberg neighbourhood, which contains the European institutions and the largest financial establishments to the east of the city, does not seem to have relaxed the market. As recalled by this manager of a Swiss bank:

> If you compare [rental charges] over the last few years, it's even worse than in Switzerland. People sent by their head offices get paid for it, but the people living here can't afford it anymore. So therefore it's a circle: you have to give them more, because they can't afford it anymore and on the same side your company supports the dramatic increase of the rental charges (June 21, 2007).

"Don't ask me to go to Kirchberg!"

Despite these high costs, our results show that the attractiveness of the CBD remains exceptionally high for the enterprises of the financial sector (Figure 2.5). On the question of knowing the localisation of surveyed enterprises if they had to move out, Kirchberg came only second, followed by Cloche d'Or (Golden Bell) in the south of the city.

These three locations together, all within the municipality of Luxembourg City, represent 77% of the stated intents expressed.

Outside the city, the locations preferred by the firms correspond to industrial estates located within the agglomeration (Bertrange, Strassen, Howald) or the close periphery near motorways (Schuttrange, Mamer). Other places in the country are hardly mentioned, in particular the second city of the country (Esch-sur-Alzette), where monumental industrial restructuring has just been launched. Not only does the choice of the CBD seem to be most popular but it is also the most exclusive. Among those enterprises that mention the CBD as their primary location choice in the case of a hypothetical move, 52% declare only that one possibility.

Figure 2.5 – What would be the best location(s) if your enterprise should move?

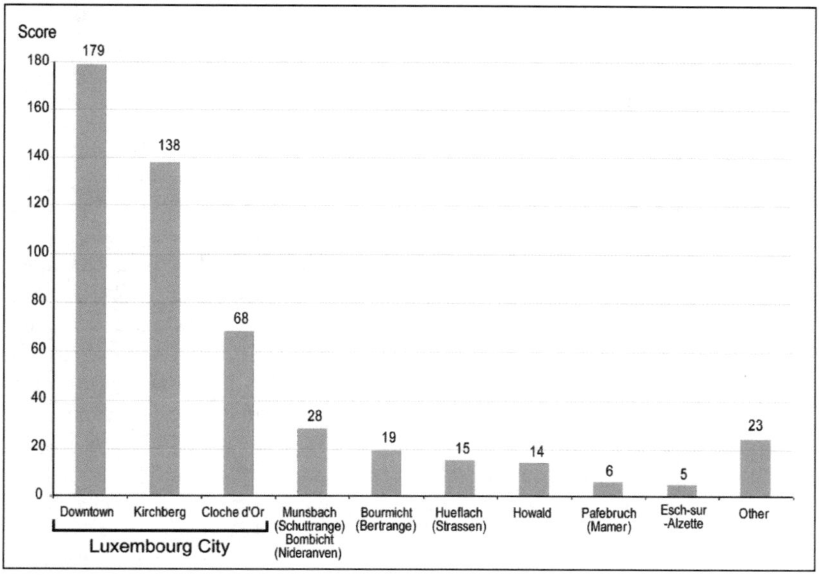

Note: N=109 firms. Source: Walther and Schulz.

Enterprises of the financial sector are subject to a broad geographical mobility that is essentially infra-urban. This characteristic was already mentioned by Nottrot (1985), who noted that 42% of banks had changed addresses between 1970 and 1979 and that 86% of such moves were within Luxembourg City. Twenty years later close to 55% of those enterprises surveyed said that they had undertaken at least one move since their installation in the Grand Duchy of Luxembourg. As in the 1980s, these moves are explained by the increasing need for office space that is larger (score = 142), less costly (34), grouped at a single location (10) and of better quality (6). Those reasons not linked to real estate appear secondary, in particular those linked to transportation such as parking space (28), congestion (8) and the reliability of public transport (5).

Enterprises are faced with the fact that most available spaces in the CBD are small. Thus, it is common to note that firms possess spaces that are disseminated across several buildings. When they have the means, firms decide to regroup their activities within a larger building, as did IBM (1991), Commerzbank (2004) and UBS (2008). Very large enterprises wishing to regroup their activities in a single building often choose the Kirchberg area. The CEO of a British bank explained the following:

> If we cannot go to Kirchberg, we would probably look at the downtown area, but the prices will probably be prohibitive for downtown. Although the prices between Kirchberg and downtown are now getting fairly similar. And then I guess Cloche d'Or would be a bit downmarket for part of our property-making business […] but it would be fine for our fund administration business, so there is always this trade-off as to which face you want to show (May 30, 2007).

Despite public transport services that are still deemed unsatisfactory, Kirchberg offers a prestigious localisation for firms wishing to extend their activities. As this partner of a London-based law firm on the verge of moving to the Kirchberg area recalled: "Basically we are looking for more space for our offices. Kirchberg is a kind of state-of-the-art location. Going to Kirchberg is also a way of improving our visibility as a law firm" (June 11, 2007). For large firms, the M&A process accelerates the need to regroup such localisations in a new building, as was the case for Mizuho (2001) and LogicaCMG (2007), which chose peripheral locations, or for certain enterprises such as PricewaterhouseCoopers and Deloitte, which moved from the periphery to the periphery depending on their expansion. Whereas large enterprises prefer Kirchberg, small ones seem to prefer the CBD where they find an environment with very dense advanced services, which is useful both for their own activities and those of their clients. As noted by this accountant located in a historic private house in the CBD:

> Don't ask me to go to Kirchberg! Normally we don't have Luxembourg-located clients, but when they come here, they can quickly go to a bank, the lawyers are here, and everything is here for the business: hotels, restaurants, so when you are there for business, the airport is nearby, and the train station (June 25, 2007).

Regarding international transport links, a different picture is painted for air, road and rail. Many interviewees consider the proximity to the airport a great advantage of a Luxembourg location but usually complain about the lack of intercontinental connections from the national airport. Business people willing to attend a meeting in New York, for example, will have to take a flight to Frankfurt or another major European hub before crossing the Atlantic Ocean. Furthermore, even though Luxair, the national airline, is a well-known business carrier that serves 50 destinations in Europe, North Africa and the Middle East, many bankers would appreciate more frequent flights between European capitals to be able to attend a full-day meeting in Zurich, for example, and return the same day. The construction of a new airport terminal in 2008 was expected to increase the number of business flights from Luxembourg. As far as highway connections are concerned, Luxembourg is considered well connected with other European cities, in particular Brussels and Germany. New TGV lines have also increased accessibility at least westwards, since Luxembourg is now only two hours from Paris.

The central place of banks within the finance industry

One of the particularities of Luxembourg is that, contrary to more diversified IFCs, most professional relationships take place within banks and towards banks. This specific aspect of the Luxembourg IFC is illustrated in Figure 6. This figure represents the orientation and intensity of the relationships within the financial sector between the various sub-sectors such as banks, investment banking, fund and asset management, insurance companies and PFSs on the one hand, and between those sub-sectors and other connected activities such as IT, law firms, auditing, accounting and other firms on the other. The measurement of those ties was obtained by asking each surveyed firm the types of firms with which they entertained the most intense work relationships on a scale ranging from 1 to 3. Figure 2.6 indicates both the direction (arrows) and intensity (thickness) of the relationship between types of firms within the financial sector (in grey) and outside the sector.

This analysis confirms the central place of banks within the financial sector. All types of enterprises state that their most important relationships are with banks, particularly legal services firms (47% of ties) and audit firms (43%), and then with enterprises in the same sector. PFSs also assert particularly strong relationships with fund and asset management and legal services. This general principle confirms the dependence of advanced services enterprises on the banking sector.

Figure 2.6 – Which types of firms do you have the most important inter-relationships with?

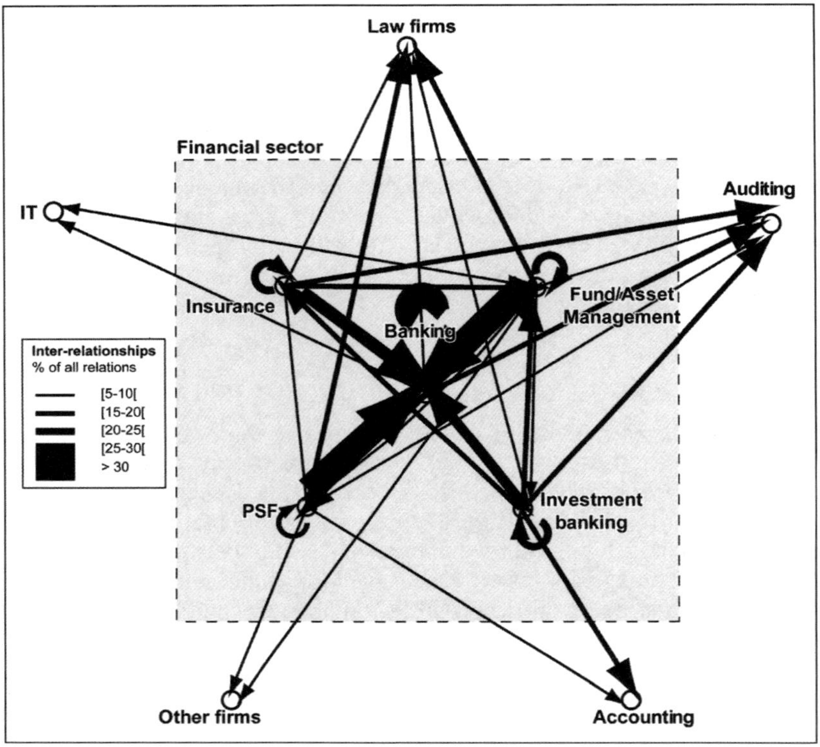

Note: Inter-relationships from outside the financial sector (in grey) are not represented.
Source: Walther and Schulz.

These results show both interesting similarities and significant differences to what Taylor *et al.* (2003) observed in their financial services clustering study in London using the same methodology. On one side, banks are located at the hub of the cluster and a very high incidence of relationships with firms from the same sector can be observed in London and Luxembourg, especially in the banking sector. On the other side, the Luxembourg cluster seems to be less oriented towards non-banking activities, such as IT, law or audit firms, than the London cluster.

The difference between the two can be explained by the absolute size of the British centre – 325,000 employees in 2006 and about 450 banks (Roberts 2008) – as well as their structural differences. With New York, London is one of the two "mega" command centres of the world city network (Taylor 2004), whose financial power relies on four main pillars: foreign exchange, international equities, exchange-traded derivatives and insurance (Budd 1999). This privileged situation is the result of the deregulation measures taken in the British financial system from the 1980s, in particular the possibility for foreign banks to access the London Stock Exchange. As a leading financial centre with a long history of trade and banking, London offers a much broader diversity of financial and support services than Luxembourg, whose prosperity primarily comes from specialised financial activities.

Conclusion: "Pick up the phone and just call him"

Even though the origins of Luxembourg's financial centre are strongly grounded in an ongoing and highly adaptable niche policy, the role of the latter is being increasingly replaced by other locational advantages. These comprise both endogenous agglomeration advantages and Luxembourg's strong embeddedness within international networks. Despite some centrifugal tendencies within the Grand Duchy, the industry remains highly concentrated in Luxembourg City's most central and/or business-oriented areas. Recent relocations towards more peripheral locations are mainly part of intra-firm division of labour between front and back office activities.

Apart from its high horizontal and vertical differentiation, the cluster's success seems to predominantly rely on its institutional and external dimensions. However, the external dimension has to be redefined given the strong influence of banking headquarters, most of which are located in Frankfurt and are usually in charge of the development of new products or tools. Local innovativeness can be found outside the actual conception of financial products and management tools, but merely in the process of making these products viable from a regulatory perspective.

As numerous interview partners have highlighted, the usually constructive interaction with national authorities, as well as the multilingual and multicultural skills of the local workforce, are a competitive advantage over other EU member states. This refers again to the institutional dimension of the cluster, which is not limited to formal and regulatory aspects. There is growing empirical evidence that the cluster has generated its own "local buzz" and strong informal interactions with local subsidiaries of multinational groups. This is also true for the relationships between firms and public authorities, as one senior manager of a Swiss bank pointed out: "The informal contacts in Luxembourg are probably the most important thing that you can have here, because you have quite a good relationship with a lot of people" (June 21, 2007).

As a financial centre, Luxembourg is a small world where everyone knows everyone. As recalled by the same interviewee: "I see all these chairmen of the private banking associations over there, so they have huge experiences and if you really have a problem, also on the government side, pick up the phone and just call him. So, it's the most important thing: informal contact and networking" (June 21, 2007). Hence, Luxembourg cannot be understood as a localised sub-cluster of Frankfurt or London, but rather as a specific local institutional setting, with established adaptation and communication strategies that deserve further scientific attention.

Note

The authors wish to thank Britta Klagge for her useful comments on an earlier version of this paper, Michaela Gensheimer for conducting the personal interviews on behalf of the University of Luxembourg in summer 2007, and the Committee on Financial Sector Supervision (Commission de surveillance du secteur financier, CSSF) for providing useful data.

CHAPTER 3

"Not always as sexy as New York..."

International Perceptions and Representations of Luxembourg City

Ralf BLÄSER, Michaela GENSHEIMER and Christian SCHULZ

This chapter explores the international perception of Luxembourg City. Based on qualitative interviews conducted with representatives of international banks and accountancy and consultancy firms based in Luxembourg, as well as on a content analysis of international print media, various images and representations have been identified. These representations go beyond standard stereotypes and allow a differentiated portrayal of how Luxembourg is perceived in the international business realm.

Complementing a more standardised survey among knowledge-intensive service firms in Luxembourg that dealt with their locational preferences and inter-firm networks (see below), a series of qualitative interviews were conducted with representatives of international firms located in Luxembourg. These interviews gained a thorough understanding of how Luxembourg as a city, business location and country is perceived from abroad. The underlying assumption of this contribution is that both the perspectives communicated by international key actors in the business sector, as well as the media coverage of a topic (here "Luxembourg"), contributes to the way a city is represented and perceived in the public realm: these discursive practices contribute to the construction of Luxembourgish realities.

The intention of the analysis of the international press was to track down how the common images and realities of Luxembourg are reproduced, modified or deconstructed through these textual elements. The qualitative content analysis of the material draws on terminology and key concepts from the discourse analysis. Nevertheless, owing to the limited scope of the study, a pragmatic approach was chosen. Neither specific linguistic patterns nor the relationship between language and the social and cultural contexts could be systematically conceptualised and examined (for the theory and method of discourse analysis see Phillips and Jorgensen 2002, Paltridge 2006, Keller *et al.* 2001). The material was instead structured along a number of dominant framings of the topic "Luxembourg" as they were identified during the course of the analysis. Put simply, a framing can be understood as the process (and/or the result thereof) whereby a topic is lifted beyond a cognitive level of meaning. Each of these framings in turn consists of various discourse threads or storylines substantiating its arguments. The tangible textual statements identified in the material are referred to as discourse fragments. They include textual arguments and rhetorical means (semantic fields), both supporting the expressiveness of each individual discourse thread (Jäger 1999).

The remainder of this paper unfolds in four parts. First, some methodological aspects will be clarified. The two main parts will then present the outcome of the two empirical studies before a final synthesis draws a variety of conclusions from the presented data.

Methodological aspects

The personal interviews with representatives of international companies located in Luxembourg City were prepared and conducted from mid-May until the end of June 2007. Interviews were structured in the sense that they followed a questionnaire including both quantitative and qualitative questions. The quantitative part of the survey was based on a similar study conducted by *The Corporation of London* in 2003 on the financial services cluster and its significance for London (Taylor *et al.* 2003). The results are presented by Walther and Schulz in this book (see Chapter 2). The quantitative part of the questionnaire had a clear focus on "hard" location factors, whereas the qualitative part was supposed to gather information and impressions regarding the following aspects:

- The perception of Luxembourg City as a business location.
- The consideration of relocating the business to be closer to cooperating companies.
- The relevance of informal contact with staff from other companies.
- The recent development of Luxembourg City as a business location.
- The perception of the multicultural diversity of the city's labour force (asset or problem?).
- The quality of life in Luxembourg City.
- The internal and external image of Luxembourg City.

Concluding the interview, people were asked for a short description of Luxembourg City in their own words, simply saying what popped into their minds. This chapter will mainly focus on the image-related sections of the interview.

In a preparatory phase, interview partners at the individual companies were identified. The list of companies to be included in the survey initially consisted of 48 global service firms[1]. Seven smaller companies (independent local companies that were members of a large international network) were not used in the survey. Two others in the insurance sector did not exist anymore. Altogether, 39 companies were eventually included and contacted, out of which 21 were interviewed.

Interview partners were usually representatives of the company's executive board, partners within the company, or heads of human resources. Most interviews took 40-50 minutes. In a few cases, interviewees had already prepared the questionnaire sent to them in advance, upon agreement of an interview date. In these cases, the interview was more focused on the open questions where people could speak freely. Two interviews were conducted by phone and were comparatively shorter than the personal interviews. In general, the interviews took place in a friendly and relaxed atmosphere. Nine of the interviewees were from Luxembourg, whereas the others were of different nationalities. In terms of critical statements (e.g. regarding expatriates' difficulties in mingling with locals), Luxembourg people often shared their international colleagues' opinions.

The print media analysis was based on a sample of 122 articles selected from a wide range of newspapers and journals from around the world. Owing to the linguistic competencies of the author, the search was restricted to German, French and English-language publications. In total, 615 English, 25 German and 10 French language papers were systematically scrutinised. The search was confined to the time period between January 2000 and July 2008. The selection of these publications was based on the knowledge and judgement of the author. Local newspapers from the Greater Region were deliberately excluded from the analysis because of their expected bias towards Luxembourg.

Because the word "Luxembourg" appears in numerous articles and news messages concerning the European Union (e.g. decisions made by the Luxembourg-based European Court of Justice) or the finance sector, the keyword search was – whenever possible – restricted to the article's headlines and introductory text. Keywords included "Luxembourg", "Luxembourgish", "Capital of Culture'" and "Greater Region" (in all three languages). This extensive search resulted in 71 German, 26 French and 25 English language articles.

---

1 Altogether, 48 out of the 100 global service firms identified by Taylor (2004) were present in Luxembourg when the survey started.

The majority of the selected articles appeared in regular newspapers and news magazines (Table 3.1). However, a few specialist papers and journals were also included in the database to increase the number of relevant articles to be examined (e.g. *Börsen-Zeitung*, *Travel Weekly*). In a next step, the text corpus was further reduced by eliminating articles that did not contain adequate text passages. Thus, 14 of the original 136 articles were deleted from the database. The sample by no means claimed to be representative of the international press at large. Rather, it was based on the availability of online databases and the subjective selection of the author.

Table 3.1 – Publications and respective number of articles in the sample

German language		English language		French language	
Frankfurter Allgemeine Zeitung	10	*The Guardian*	8	*Le Nouvel Observateur*	1
Berliner Zeitung	3	*South China Morning Post*	5	*Le Temps*	3
Börsen-Zeitung	12	*Sunday Tasmania*	1	*Les Échos*	6
Handelsblatt	10	*Banker*	1	*Le Point*	3
Die Welt	7	*Sunday Telegraph Australia*	1	*La Tribune*	3
Süddeutsche Zeitung	5	*Travel Weekly*	1	*Le Figaro*	3
Frankfurter Rundschau	5	*Daily Yomiuri*	1	*Le Monde*	7
Welt online	3	*International Herald Tribune*	1		
Tageszeitung	3	*The Financial Times*	1		
Welt am Sonntag	2	*The New York Times*	1		
Kölner Stadt-Anzeiger	2	*The Boston Globe*	1		
Der Standard	2	*Prague Tribune*	1		
Die Zeit	2	*Financial Times Expat*	1		
Financial Times Deutschland	2	*Conference and Incentive Travel*	1		
Spiegel Online	2				
Total	**71**		**25**		**26**

Sources: authors' survey.

The vast majority of the articles appeared during 2006 and 2007 (Figure 3.1). This clustering was caused by the fact that Luxembourg and the Greater Region held the title European Capital of Culture in 2007. Together with other discursive events around this time – such as the inauguration of the architectural landmark the Museum of Modern Art (Mudam) and the Philharmonie – this led to increased media attention. These occasions are also the reasons why the majority of articles in the sample cover issues related to recent cultural developments in Luxembourg, followed by financial topics (Figure 3.2).

Figure 3.1 – Temporal distribution of the sample

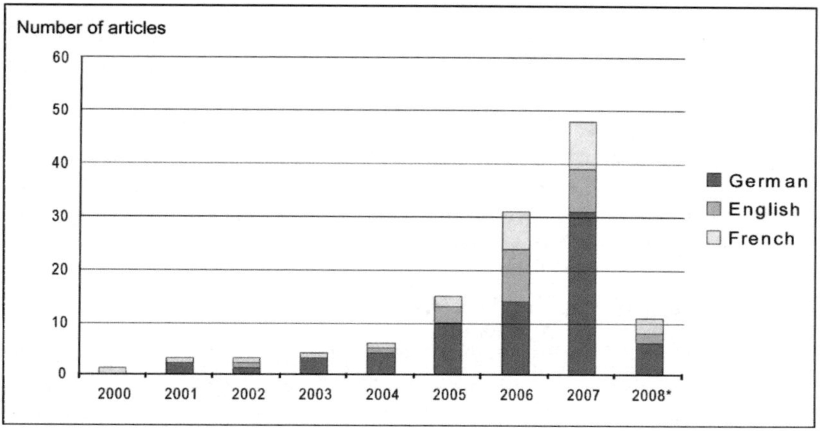

Sources: authors' survey.

Note: * = as to July 2008.

Figure 3.2 – Thematic priorities of the articles

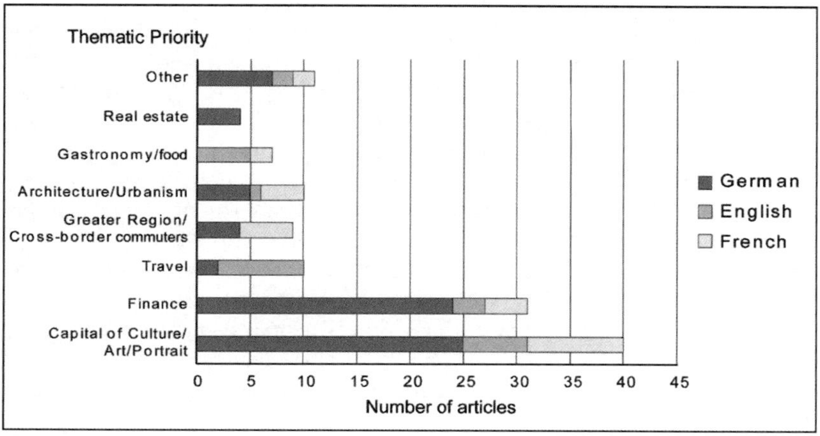

Sources: authors' survey.

In the first analytical step, dominant framings and their respective storylines were identified from a random sample of about 20% of the articles. These were then used to develop a comprehensive coding system to structure the material.

Obviously, there was a risk that the views represented in the selected articles would be biased because of publicity campaigns and other semi-official publications from Luxembourg; these are sometimes deliberately adapted or rewritten by foreign journalists. Hence it has not always been easy to distinguish between presumably "authentic" statements (often the more critical or ironic ones) and the merely copied information and clichés. Most of the quotations used below mirror at least a minimum degree of independent investigation and own inquiry into the particularities of the country or the city. However, from a discourse analysis perspective, this potential bias is not of primary importance since both the "authentic" and the "copied" views disseminated by the print media are parts of the international discourse around Luxembourg – and as such a constitutive element of the general perception.

Luxembourg in the eyes of international corporations

The following section presents the most important findings from the corporate interviews. Owing to the summarising character of this report, the number of quotations was limited to a selection of key statements.

The perception of Luxembourg City as a business location

The overall perception of Luxembourg City as a business location is rather positive, particularly because of its attractive legal and physical environment. As one interviewee put it: "Authorities are quick to react… which makes us less administrative than other countries." Even though this perception is not shared by all interviewees, the Grand Duchy's capital is seen as an entrepreneurial city where it is easy to do business. Its comparatively small size facilitates access to relevant services for business clients and partners. In terms of banking, particularly fund administration, a Luxembourg location is considered a quality label. It is appreciated for its flexibility in both financial and labour market regulations. One example is the introduction of night employment at one bank to better cooperate with business partners/clients in other time zones. One interview partner pointed out that Luxembourg was much more result-oriented than other business locations he knew. By contrast, Luxembourg as a country – and consequently the city of Luxembourg – is perceived as conservative and "not as open-minded as it could be".

Because many business activities in the city deal with back and middle office tasks, they could be subject to the threat of outsourcing to other countries, e.g. India. For some companies the only reason they are located in Luxembourg is because their clients are there. Statements such as "if there was no market we would not be here" were only made by representatives of firms offering support services to the banking sector, i.e. accountancy and consultancy firms. One company representative pointed out that "the informal contacts in Luxembourg are probably the most important thing that you can have here". They are seen as a "mine of information" since news spreads quickly this way. Therefore, they help stay up to date with developments. According to one interviewee, it is "important that even competitors keep in contact informally to make sure that everyone goes in the right direction".

The recent development of Luxembourg City as a business location

The recent development of Luxembourg City as a business location is assessed rather favourably. In fact, most interviewees observed a certain evolution in different fields. In asset management, administrative platforms have been and will continue to be concentrated in Luxembourg. Regarding the labour market, which has traditionally been comparatively multicultural, an increased internationalisation has been observed. Although employees in the financial sector used to come from the different areas of the Greater Region, they now come from more distant areas in France (e.g. Toulouse, Nantes), Belgium (e.g. Antwerp, Brussels) and other European countries. According to one interviewee, this aspect contributed to an "increased professionalism" because these people are usually young, willing to learn and work hard to build a career.

Various interviewees appreciated the efforts being placed on upgrading the cultural life of the city, e.g. the Philharmonie, the Coque (the national centre for sports and culture), the creation of the Mudam and the renovation of existing museums (see below). There is an impression that development is "very well organised" by the government, which, according to one interviewee, had not always been the case. As a business location, the city is "showing that it's resilient and flexible and can adapt to a changing environment".

Nevertheless, the overall positive development places a lot of requirements on the local labour supply and infrastructure. As one interview partner expressed it: "There is a big upsurge in business but then there is a big shortage of qualified staff." Several of the interviewed companies go abroad to take part in recruitment events and contact universities to hire people with an adequate profile. However, they are also confronted with obstacles in this respect. Some interviewees pointed out that it is not possible to hire people from eastern European countries or India due to work permit restrictions. Another hampering factor can be the fact that Luxembourg is "not always as sexy as New York or other financial centres" when recruiting from abroad. Moreover, when it comes to employees of Luxembourgish nationality – who could be interesting because of their language skills – the private sector has to compete with the public sector, where the majority of Luxembourg people seek employment because of the considerably higher remuneration.

The perception of the multicultural diversity of the city's labour force

One of the particularities of the Luxembourg labour market is its multicultural diversity. A large share of employees in the financial sector comes from the border regions around the Grand Duchy in Lorraine, Wallonia, Rhineland-Palatinate and Saarland.

All interviewees except one considered the multicultural diversity of the labour force to be one of the city's most important assets. It is perceived as a "richness" that makes the city a "field study of Europe", which in turn is conducive to attracting young educated people to Luxembourg. The multiculturalism of the labour market does not only convey that "everybody can work here". It is an essential prerequisite to doing business in Luxembourg because the country's internal market is relatively small and its economic well-being depends on the demand of services from abroad. To adequately meet this demand, staff with language skills, as well as different cultural backgrounds, are needed. In this respect, the need to encourage work permits for staff from eastern European or non-EU countries was re-emphasised.

When asked whether this multiculturalism has ever caused problems for a company, the majority of interviewees answered with a straightforward "no". Some representatives of bigger companies mentioned that their staff consisted of between 20-30 nationalities with no problems. As one interviewee pointed out: "It's not a question of being of a different nationality, it's a question of human relations, it's how people are open-minded." A few people conceded that there might be some "small problems" or "challenges", most related to language issues. These can become a difficulty for companies with a predominantly francophone staff that might have to deal with German clients or the other way around, because many Germans do not speak French, and many francophones do not speak German.

The representative of one company explained that he was recently surprised to be confronted with such language obstacles. After his company had hired a few German employees in addition to the predominantly French-speaking staff, they had to adapt their internal training courses accordingly and switch their course language from French to English. Even though English seemed to be the common denominator for all, not everyone had a sufficiently good command of the language to benefit fully from the course. The ideal employee for many companies would be a Luxembourger who is fluent in at least three languages. However, as pointed out before, Luxembourg people are rarely interested in this job market.

The quality of life in Luxembourg City

The quality of life in Luxembourg City is generally viewed as very high. Interviewees appreciate the cleanliness, neatness, short distances and, above all, feeling of security, which was mentioned comparatively often. In fact, one interviewee identified the "safety of living" as the "major merit" of the city, which he considered "a very good place for a human being to live".

Despite being relatively small, most people interviewed believe that Luxembourg City offers enough diversity and has an enormous amount of cultural activity. It is thought to be good for family life and an ideal place to raise children. Another asset is its close proximity to nature. Finally, the central location of the city in Europe makes it easy to reach several other European cities within a comparatively short period of time. One representative of a company particularly enjoyed the multicultural atmosphere in the city and the chance of meeting people from different backgrounds, giving everyone the opportunity to find someone to identify with in terms of language, culture and personal values.

The downside of the city is its lack of vibrancy. As shops close at six o'clock in the evening, there is usually little post-work activity in the city centre. The city is perceived to be "dead in the evenings" and "too calm". Extended opening hours would be appreciated by the people interviewed. Also, there is not enough entertainment at the weekends. One interviewee described the city as basically a "ville de province" in French terms, whereas another admitted that although he liked Luxembourg City he could not stand spending every weekend there.

Even though most people were satisfied with the quality of life in Luxembourg City, they supposed that the social life for young people might be unattractive. "It's good for work, it's good to be here in the week and you can do something, but Luxembourg City is quite lonely at the weekends when businesses are closed."

Some believe that the establishment of Luxembourg University could contribute to bringing more life and more alternatives for young people. One interviewee suggested that the university could offer events at weekends with a certain intellectual appeal, targeting young professionals that would usually seek diversion and amusement in neighbouring regions across the border.

Rental costs were another aspect criticised by some. Owing to the high rent for flats and houses, many people commute across the border every day, which also partially accounts for the quiet streets in the evening. Rents for retail space in the city centre are equally considered to be too expensive. They represent an obstacle for business people who might otherwise be interested in opening a shop, thereby also amplifying the range of shops. This is seen as a reason for not having a wider choice of "alternative" shops in Luxembourg.

The image of Luxembourg City

It is not always easy for interviewees to distinguish between an internal and external image of the city. Very often, the aspects mentioned before were repeated, e.g. with respect to quality of life or the perception of the city as a business location. As for the internal image of Luxembourg City, there was a tendency to distinguish between the business and private side.

On the business side, the city has a positive image because of the existing regulatory environment, flexibility and the professional and solution-oriented attitude of stakeholders. Regarding private life, however, the society seems to be split in two, with Luxembourg-born people on the one side and expatriates on the other. A common observation of all nationalities is that international people moving to Luxembourg have serious difficulties integrating into the national community. Instead, they often have an international circle of friends. One interview partner, who is a native of Luxembourg, explained that this could be because of the Luxembourg culture "where from an early age on you have a restricted circle of friends which is difficult to break after a certain age".

From an external point of view, Luxembourg City's image seems to be characterised by a lack of information, which can either create the wrong image or no image at all. Outside the Grand Duchy, people often seem to associate Luxembourg City with a "tax paradise" or just see it as a "banking centre". A lot of people do not even know Luxembourg, nor could they spot it on a map. At best, "Luxembourg is seen as a small village" by foreigners "even if we [the Luxembourg people] think it is a big city". Because of this lack of knowledge, one representative of an international bank reported that he had been asked how he "ended up in Luxembourg". Apparently, a lot of people think that Luxembourg is "a boring place where there is nothing to do and people just work and do banking".

However, when people come to Luxembourg for the first time they see a lot of positive aspects and consider it a nice place. As a consequence, the most important effort, according to one interviewee, should be to convince people to think about Luxembourg and to properly communicate what the city has to offer, e.g. in cultural terms. Communication and marketing, though, are perceived as weak and insufficiently developed. In this context, the city's initiative to hire a city manager was received positively by some interviewees.

Luxembourg City in a nutshell

At the end of each interview, the respondents were asked to provide a short statement of how they would characterise Luxembourg City in a nutshell. Here, it was often described as "comfortable", "cosy", "beautiful" and "safe". It is seen as "well organised", "a great place to work" and "to spend money". Further statements include:

> A small and nice cosmopolitan city in the European Union.
> Small, interesting, multicultural, but unknown.

The comparatively limited size of the metropolis was used several times to describe but also distinguish it from other international centres such as Paris, Brussels or London. Furthermore, the multicultural population was identified as an obvious characteristic of Luxembourg City. Some pointed again to the difficulty of mingling with local people.

One interviewee described the city as "socially rather upscale" with "no huge discrepancies between rich and poor", as is usually found in other international cities. Others added to this aspect by depicting the city as being "exclusive" and "expensive", particularly in terms of real estate. Finally, in spite of the mainly positive descriptions, it was also mentioned that "there is a lot of space for improvements regarding the appreciation across the border", again underlining the importance of better marketing outside the Grand Duchy.

Luxembourg as seen by the international print media

Here, six dominant framings were identified: "multicultural society"; "place of high quality of life"; "nation of wealth and prosperity"; "small but mighty centre of arts and culture"; "major financial hub"; and "transborder integrator par excellence". Moreover, two "meta-narratives" across the lines of these framings were distinguished for further analysis: "micro-state in transition" and "strange land of uniqueness and contrast". In what follows, each individual framing and meta-narrative will be discussed and substantiated with its respective storylines and discourse fragments.

Dominant framing 1: Multicultural society

One of the most prevalent framings identified in the international press refers to Luxembourg as a multicultural society, even though only two headlines make explicit reference to this framing ("Diversity in the heart of Europe", *South China Morning Post*; "Multikulti im Zwergstaat", *Die Zeit*). Overall, 75 textual passages bear this allegedly distinctive feature of the Grand Duchy. The main purport of the fragments is that "la question de la multiculturalité est intimement présente dans la vie quotidienne du Grand-Duché", as the Suisse newspaper *Le Temps* suggests. Other articles take the same line and conceive Luxembourg as a "unglaublich verdichteter Migrationsfluchtpunkt" (*Frankfurter Allgemeine Zeitung*) or even as the "Schmelztiegel Europas" (*Berliner Zeitung*). Two prime discursive threads support this framing: immigration and multilingualism.

Overall, a broadly affirmative tone resonates in most accounts on immigration. Frequently, the high percentage of migrants in the Luxembourgish population (albeit using varying figures) is referred to as a starting point to the topic, often with a somewhat impressed or even admiring undertone. Regarding the integration of migrants into the so-called host society, a discourse of harmony prevails. The *Süddeutsche Zeitung* even asserts that the high percentage of migrants has never caused any hostilities ("ohne dass das je zu Feindseligkeiten geführt hätte"). "The well-integrated foreign residents" (*South China Morning Post*) are in many cases considered a valuable asset for Luxembourg, as the following three quotes exemplify:

> Der Austausch der Kulturen ist Teil des Alltags und wird seit je als Bereicherung verstanden. (*Welt Online*)
>
> Aus dem kulturellen Miteinander aber bezieht es Kraft. (*FAZ*)
>
> Ohne die Ausländer wäre die Stadt nicht, was sie ist: eine sehr kleine und zugleich sehr kosmopolitische Metropole. (*Spiegel Online*)

This putative story of harmony is rarely explained or even critically scrutinised. If at all, the absence of conflict is routinely ascribed to the nation's prosperity, allowing for a relatively high standard of living for all segments of the population. Only one article, in *Welt Online*, offers a rudimentally critical explanation:

> Die Überschaubarkeit des Luxemburgischen Migrations und Integrationsproblems lässt dabei manchmal den Anschein einer Idylle aufkommen. In Luxemburg ist man nun einmal sehr weit weg vom Mittelmeer und der Ostgrenze der Europäischen Union. Es ist eben doch ein glücklicher Zwergstaat, in dem Westeuropäer weitgehend unter sich bleiben.

Only three articles in the sample make reference to a potential problem arising from the multicultural composition of Luxembourg's society, namely the fear of losing their identities ("gewisse Entfremdungssyndrome", *Frankfurter Allgemeine Zeitung*) – especially their native language Letzebuergesch – among segments of the elderly population.

A concrete contribution to these rather positive accounts on immigration is a discourse thread relating to the particular atmosphere and *savoir vivre* in Luxembourg City that is framed to be caused by this "migrant culture":

More than half of its inhabitants are foreigners – most notably Italians, Spaniards and (particularly) Portuguese – and these recent arrivals give the city an almost Mediterranean air. (*The Guardian*)

Occasionally, the pervasive image of Luxembourg as the ideal blueprint for Europe also draws upon the multiculturalism discourse, as the following textual passage shows:

Als Versuchsfeld begreift man sich deshalb in dieser hübschen Miniaturausgabe Europas, für den Strukturwandel – von der Schwerindustrie zur Finanzwirtschaft – wie für das, was man multikulturelle Gesellschaft nennt. (*Süddeutsche Zeitung*)

With regard to the linguistic competences of Luxembourgers, the majority of the press articles examined draw a rather uncritical picture of an entirely multilingual population in which almost everyone speaks three or four languages:

This means most locals are multi-lingual. Luxembourg has its own national language, known as Letzebuergesch, but almost everyone also speaks French and German, and most in service industries have a good command of English. (*The Guardian*)

Das Sprachengewirr, das auf Straßen und in Geschäften herrscht, erscheint umgekehrt babylonisch – alle verstehen sich immer und überall. (*Welt Online*)

In conclusion, the international press broadly – and rather uncritically – frames Luxembourg as a multicultural and multilingual society in which associated problems, such as linguistic discrimination or racism, rarely exist. This feature alone makes the Grand Duchy an interesting and uniquely cosmopolitan place that could act as a model for the state of the future, as it "stands as an eloquent example of the integrative strength of a growth-based economy with social cohesion and multilingualism at the core of its national identity" (*South China Morning Post*).

Dominant framing 2: Place of a high quality of life

Luxembourg, especially the city, is frequently portrayed as a cosy place with a high quality of life, where all the wants of a sophisticated lifestyle are catered for. This ranges from a plethora of culinary delights to an attractive natural and cultural landscape, where one cannot only breathe fresh air but also smell the scent of history.

In general terms, many accounts stress the fact that Luxembourg is a city where "l'art de vie est porté à un rare degré d'excellence", as the France's *Le Figaro* notes. High-class cuisine and other culinary delights such as confectionaries and the Moselle wine are frequently mentioned as Luxembourg's prime assets when it comes to quality of life. Here, not only the unusually large number of Michelin star restaurants in the city but also the typical combination of French quality and German quantity is pointed out, as these quotes illustrate:

> Fine dining is a Luxembourg specialty. It boasts more Michelin-hat restaurants per capita than any other European capital. Its cuisine is a mix of French finesse and German heartiness. (*Sunday Telegraph Australia*)

> The people of Luxembourg take their bakeries and patisseries seriously. (*South China Morning Post*)

In one albeit exceptional article, the high quality cuisine argument is supported by a discourse fragment that stresses the allegedly untouched Luxembourgish nature:

> [...] un grand de la gastronomie qui s'ignore ou, plutôt, qui oublie de le faire savoir. Car, là-bas, depuis plusieurs décennies, l'écologie, la santé, le bien-être des animaux, tout comme l'esthétique des conditionnements font partie du quotidien [...] Il possède cependant tous les ingrédients nécessaires à une gastronomie variée: rivières poissonneuses, forêts giboyeuses, terres à vignobles. (*Le Figaro*)

Another interconnected discourse thread relates to the urban lifestyle and ambience in Luxembourg City. The common purport of most articles is that Luxembourg has plenty to offer in this respect, certainly more than one would expect in "a place better known for banking and commerce than Bohemian-style nightlife" (*Boston Globe*). Hence, an undertone of positive surprise resonates in some accounts on this topic:

> Buried in the ravine below is Grund – an idyllic village in the city centre, beneath the city wall. This was where the poor people used to live, outside the protection of the ramparts, but lately it's been spruced up a bit and today it's charming. I was afraid I'd find a yuppie theme park – the sort of smart but soulless place where wealthy bankers park their Porsches – but I'm glad to say I was mistaken. In between the quaint old cottages are plenty of intriguing restaurants and galleries. (*The Guardian*)

> The scene greeting us at Cafe des Artistes could have been in Paris, Berlin, Amsterdam, or any number of European hot spots. In a dark, smoky room, a chanteuse, young and lovely and with a world-weary voice, sat perched on a high stool by a piano. The requisite cigarette in hand, she warbled the timeless refrain of "La Vie en Rose" to an appreciative crowd who piped in during the refrain. (*Boston Globe*)

Only one article points out that Luxembourg is not exactly a place of boisterous youth culture where unconventional behaviour and fashion styles are ubiquitous:

> Und man ist hier ein bisschen spießiger als in anderen Städten – ausgeflipptes Äußeres ist nicht gern gesehen in Luxemburg, selbst wenn man

Popmusik moderiert [...] In der Tat, Jugendliche mit verrückten Klamotten oder bunten Frisuren sucht man in Luxemburg vergebens. (*Spiegel Online*)

Regarding the capital's quality for a good stroll, only the Vienna daily *Der Standard* makes a rather critical observation: "Es gehört keineswegs zu den Stärken des Großherzogtums, dem Flaneur ein Kaffeehaus oder eine Bar zur Seite zu stellen." Other accounts, on the contrary, appreciate the city's capability to meet the needs of the flâneur, as this excerpt from *Spiegel Online* exemplifies: "Kleine Geschäfte, quirlige Straßencafés und noble Restaurants verlocken zum Verweilen und Einkaufen."

Other discursive aspects in the international press that add to Luxembourg's image as a locus of a high quality of life are its contemplativeness, in concert with its history-laden picturesque urban landscape. Here, the city is frequently represented as an almost fairy tale-like romantic place in which centuries of European history can be experienced firsthand. In particular, fortress-related semantic fields such as the commonly used metaphor of the "Gibraltar of the North" are employed to emphasise Luxembourg's historical uniqueness in northwest Europe:

> Von den vielen Spielarten der Kunst wurde hier, im "Gibraltar des Nordens", jahrhundertelang vor allem eine gepflegt: die Festungsbaukunst. Auf diesem Gebiet hat man wirklich Bedeutendes geleistet – unübersehbar, wenn man sich der Stadt Luxemburg nähert. Noch heute muten die längst geschliffenen Überreste der alten Festungsanlagen so an, als wären sie in mythischen Zeiten aus dem Gestein des Bockfelsens herausgewachsen. Hoch über dem Tal der Alzette gelegen, von tiefen Schluchten umgeben, von bis zu drei massiven Ringmauern auf einmal geschützt, war eine solche Feste uneinnehmbar. (*Süddeutsche Zeitung*)

> La sévérité de l'architecture militaire a comme fondu dans la verdure des jardins en terrasses. Les vestiges des fortifications ont la mélancolie de ruines romantiques – elles plaisaient tant à Victor Hugo et inspirèrent à Turner de magnifiques estampes de la ville. (*Les Échos*)

Furthermore, the natural landscape in which this "ungemein pittoreske Städtchen Luxemburg" (*Der Standard*) is embedded is conceived to increase the city's cosiness. Characterisations such as "Finanzmetropole im Modellbahn-Look" or "Märklin-Stadt" (*Spiegel Online*) illustrate the effort to depict Luxembourg as a manageable, small-scale, but also staidly place that offers a high quality of life. The following quote supports this argument:

> Storybook is the word that comes to mind when taking in the city, whose walkable Old Town of stately painted buildings and narrow, winding cobblestone streets commands a dramatic perch overlooking scenic valleys and gorges. (*Boston Globe*)

Overall, Luxembourg is characterised as a rather placid place that provides all the amenities for the affluent and well-settled middle class. Although some of its nightlife features are framed to be somewhat metropolitan, the general thrust of the articles is one of cultivated contemplativeness. Although the demands of a sophisticated clientele are well catered for in an exceptionally picturesque setting, adherents of alternative or bohemian lifestyles would feel rather out of place in the Grand Duchy's capital.

Dominant framing 3: Nation of wealth and prosperity

One of the more classical images of Luxembourg, which is also prominent in the analysed press articles, is that of a wealthy nation with rapid economic growth, where social problems rarely exist. This framing is mainly substantiated through two interlinked discursive threads, one related to Luxembourg's exceptional economic performance and the other to the attractiveness of its job market for cross-border commuters from neighbouring countries.

When it comes to pure economic performance, almost all of the articles put Luxembourg ahead of other nations by citing the standard economic indicators:

> On croit souvent que la Suisse ou les États-Unis sont les pays les plus riches et les plus performants de la planète. Il n'en est rien. Ce tout petit pays de 424,000 habitants, coincé entre l'Allemagne, la Belgique et la France, caracole en tête de tous les palmarès depuis vingt ans, ignorant crises et chômage. (*Le Monde*)

> Wirtschaftlich allerdings ist das Reich Großherzogs Henri I. ein Gigant. Mit einer Wachstumsrate von 6,2 Prozent stellte es im vergangenen Jahr sogar das prosperierende Irland in den Schatten. (*Die Welt*)

The second thread deals with a straightforward side effect of this wealth: the abundance of jobs, which exerts a huge pull on jobseekers from the neighbouring countries. As expected, only the French and German papers in the sample elaborate on this issue. Here, the Grand Duchy is – and arguably correctly so – represented as a "Jobparadies" (*Spiegel Online*) or even the "promised land" ("gelobtes Land", *Handelsblatt*), which provides not only for its own population but also for a large share of the employees from the structurally weak regions beyond its borders:

> Des salaires de 15 à 30% supérieurs. Et les Lorrains ne se font pas prier pour traverser la frontière! Premier avantage qu'ils tirent de leur activité professionnelle grand-ducale: le niveau de salaire. (*Le Point*)

> Kein Ort ist begehrter – für die umliegenden, strukturschwachen Regionen fungiert das Großherzogtum als Jobbörse. (*Frankfurter Rundschau*)

In conclusion, the articles characterise Luxembourg as a nation of unique wealth and economic surplus for the most part, without dealing with the potential threats to this prosperity.

Dominant framing 4: Small but mighty centre of arts and culture

In the wake of its award as European Cultural Capital 2007, culture and the arts have taken centre stage in the press coverage on Luxembourg (Figures 1 and 2). Here, the Grand Duchy is characterised as a rising star in the cultural sky over Europe, with the inauguration of the Mudam and the Philharmonie, which are mentioned as the turning point in Luxembourg's efforts to develop into a cultural metropolis. At the same time, its formerly provincial approach to culture is discussed in several pieces, thereby accentuating the novelty of the current developments. In general, Luxembourg until the mid-1990s was framed as "an artistic and intellectual backwater" (*The Guardian*), in which cultural indifference and narrow-mindedness prevailed, as these two passages suggest:

> Man kann ohne Übertreibung sagen, dass Vauban, der französische Festungsbaumeister, jahrhundertelang der mit Abstand bedeutendste Künstler war, der sich in Luxemburg, nun ja, schöpferisch betätigte. Die Kunst hatte hier nie ein Heimspiel. (*Süddeutsche Zeitung*)

> En 1995, lorsque Luxembourg fut, pour la première fois, promue "capitale européenne de la culture", il fallut, à la demande des autorités ecclésiastiques, voiler une statue de Nikki de Saint Phalle. Et quelques bourgeois s'émurent que l'on montre dans un musée les œuvres au crayon à bille bleu du Flamand Jan Fabre. Confrontés à ce pesant conservatisme, à cette bigoterie de sous-préfecture, à cette caricature du Grand-Duché par lui-même, quelques créateurs locaux faillirent abandonner ce pays minuscule à ses banques, son PIB pharaonique et ses cigarettes sous-taxées. (*Le Monde*)

The British daily *The Guardian* even offers an explanation for Luxembourg's late artistic awakening: "Well, its proximity to other countries has been an economic blessing – but for a long time it was an artistic curse. Luxembourgers looked abroad for their cultural fix, just as they went abroad to study." The press coverage of the two aforementioned cultural flagships Mudam and Philharmonie, however, is largely euphoric and sometimes even reads like a sigh of relief:

> Pour tenir son rôle de capitale culturelle, le Luxembourg s'est doté d'un nombre impressionnant de lieux culturels. Le Mudam, le Musée d'art moderne, est de ceux-là. Les cimaises du musée flambant neuf de l'architecte I.M. Pei, exigent de la présence et de l'ampleur. (*Le Temps*)

> Andere Großtaten sind der Bau der Philharmonie und der Bau eines Museums für moderne Kunst, den Architektenstar Ieoh Ming Pei besorgte. Die Hauptstadt steigt damit kulturell in die Oberliga auf. (*Süddeutsche Zeitung*)
>
> Es [das Mudam] ist der vorläufige Höhepunkt der Umgestaltung des wohlhabenden Großherzogtums von einer drögen Banken- und Bürokratenstadt zur Kulturmetropole. (*Die Welt*)

Even though these two institutions seem uncontested in their importance in Luxembourg's recent transformation into a place where culture and the arts are increasingly appreciated, other less prominent cultural institutions are also covered, albeit not in an exclusively favourable manner. The most frequently cited among these are the Abbey Neumünster, the Grand Théâtre, Le Criminel[2], with its workshops and studios and the Rockhal (music theatre for rock and pop concerts) in the city of Esch. According to one article in the *Süddeutsche Zeitung*, the National Museum of Art and History is not really geared up to support the capital's efforts to become a centre of arts, because it exhibits a "Sammlung von erschreckender Provinzialität". Interestingly, a number of pieces also refer to the internationally renowned exhibition "Family of Man" by the Luxembourger Edward Steichen in the small town of Clervaux ("eine der berühmtesten Fotoausstellungen der Welt", *Frankfurter Allgemeine Zeitung*), thereby geographically expanding Luxembourg's putative metropolitan cultural asset from the capital to the remote north of the country.

Another aspect dealt with in some of the articles is the valorisation of industrial heritage sites for cultural purposes, especially in the country's south. One potentially discursive thread in this context is conspicuous by its absence in the grand majority of the pieces: education and science. Only one article in the sample is solely dedicated to the newly founded University of Luxembourg. Titled "Multikulti im Zwergstaat", this piece in the German weekly *Die Zeit* takes a sceptical stance towards the university's ambitions of becoming a truly international institution by stressing its still overly provincial impression:

> Mit einigen Kommilitonen steht der Wirtschaftsstudent unter den Ahornbäumen auf dem Campus Limpertsberg in Luxemburg-Stadt. Von einem Leben und Denken ganz im Zeichen Europas sind sie weit entfernt. "Wir wohnen ja noch zu Hause, für uns ist das eher wie eine Verlängerung der Schulzeit." [...] Trotz Mehrsprachigkeit vieler Studiengänge, bereits bestehenden Kooperationen mit 60 Hochschulen in 20 Ländern und geplanten Partnerschaften mit Unis in China, den USA und Afrika befürchten Kritiker, dass die Uni.lu eine Provinzhochschule bleibt. Schließlich kommen immer

2 The old name of the Robert Bruch Building in the Centre Culturel de Rencontre Abbaye de Neumünster, a former military hospital and prison, nowadays used as a cultural centre.

noch vor allem Luxemburger, die von zu Hause nicht wegwollen, und junge Leute aus der Grenzregion, aus Frankreich, Deutschland und Belgien hierher zum Studieren. (*Die Zeit*)

A couple of other articles merely mention the plans to establish a "pôle universitaire ambiteux" (*Le Point*) in Esch-Belval without further elaborating on the metropolitan potential of this research and education cluster.

A further discursive thread adding to Luxembourg's image as a centre for the arts and culture is that on architecture and urbanism. Three related discourse fragments can be identified in this context. The one on fortification architecture has already been dealt with previously. In one account, it is used to argue for a historic continuity in the contemporary development of Luxembourg's architectural landscape:

> Und wieder einmal fällt der Architektur eine wichtige Rolle zu. Sind doch die wesentlichen kulturellen Prägungen Luxemburgs architektonischer Art: Alle großen Militärbaumeister haben an den legendären Kasematten gebaut, die einst 23 Kilometer lang waren und das Großherzogtum als "Gibraltar des Nordens" berühmt machten. Diese steinernen Fortifikationen bekommen nun also Konkurrenz. (*Die Welt*)

Most articles dealing with this issue point to the anti-urban character of the Kirchberg Plateau with its numerous American-style office buildings and general lack of atmosphere. The most expressive among these are two statements in the *Financial Times* of London and the German *Die Zeit*:

> But its heart is, unfortunately, entirely heartless. EU-ville is an almost caricatured paradigm of faceless bureaucracy. Worse than anything conceived by Jacques Tati in his brilliant filmic satire of the modern metropolis, Playtime, it is an area separated from the city by a ravine, void of urban life or activity, of retailing or dwelling: the exemplar of the zoned city, modernism's most disastrous big idea. (*Financial Times*)

> Doch amerikanischer kann Europa kaum aussehen als hier. Auf enteignetem Ackerland, 360 Hektar groß, ist seit den sechziger Jahren ein eurokratischer Klotz nach dem nächsten emporgewachsen, Hochhäuser für die Verwaltung des EU-Parlaments, für Gerichtshof, Eurocontrol, Rechnungshof – lauter gesichtslose Monstren, die Luxemburgs beschaulicher Altstadt bedrohlich nahe rücken. (*Die Zeit*)

Likewise, many articles claim that Luxembourg has recently developed into an internationally noted architectural course of late modernism: ("vielbeachteten Architektur-Parcours der Spätmoderne", *Süddeutsche Zeitung*; "international einen Ruf als zeitgenössisches Architekturmuseum", *Welt Online*). The advancement of "cutting edge architecture" (*The Guardian*), especially on the Kirchberg-Plateau, is qualified as a deliberate political strategy – initiated by the former Prime Minister Jacques Santer – in the manner of Mitterand's *Grands Projets* suitable for putting the Grand Duchy on the architectural world map. Overall, newly constructed solitaires such as the Mudam or Philharmonie are framed as "rays of hope" amid a rather bleak urban landscape showing the way to a brighter future:

> Cette perle (the Philharmonie, R.B.) raffinée contraste avec les mastodontes tertiaires du voisinage, oasis dans un univers de bureaux. (*Les Échos*)
>
> Pour ne pas altérer son noble caractère, les urbanistes ont imaginé sur le plateau de Kirchberg une nouvelle Luxembourg mise en scène par le fleuron de l'architecture mondiale: Bofill, Portzamparc ou Pei. (*Le Figaro*)

Although the majority of the comments on the architectural quality of the new edifices are positive, the *Financial Times* takes a rather critical opinion, at least on Pei's masterpiece:

> It is solid, bourgeois stuff, the equivalent of the grand classical monuments of the 19th century that attempted to embody nationhood as much as the enriching experience of art. And it achieves a dullness as complete as the dullest of provincial museums. The building is beautifully detailed, generous, spacious, built of rich stone and steel, yet it has nothing to say. The past 40, perhaps even the past 140 years seem to have counted for nothing in a building that would look at home in 1970s Denver or 1980s Dubai. I can think of no better argument against employing an international superstar than this slab of stony boredom. (*Financial Times*)

In summary, Luxembourg's recent endeavours to upgrade its cultural sector are mostly classified as successful. Furthermore, the resulting cultural landscape is framed as an important metropolitan asset that has already lifted the Grand Duchy from a once provincial backwater onto the international stage. At the same time, this success story largely remains restricted to what has been labelled mainstream culture. Only five textual passages in the entire material refer to alternative or subcultural activities and institutions. If there is anything that constitutes a real cultural metropolis, it is cultural diversity, not only in terms of artistic styles and modes of representation but also with regard to production and consumption environments – an aspect that is almost completely missing in the analysed material and arguably also in the Grand Duchy of Luxembourg.

Dominant framing 5: Major financial hub

One of the most commonly reproduced images in the public discourse on Luxembourg is that of an important banking centre. It is, therefore, not surprising that this framing is also prominent in the press articles reviewed. Although the reporting on this issue is consistently positive in all articles in the sample, those taken from specialised business and financial papers (e.g. *The Banker*, *Financial Times*, *Handelsblatt*) deal with Luxembourg's respective competences and comparative advantages in greater detail. Here, the Grand Duchy is frequently portrayed as a highly innovative "Know-how-Standort" (*Börsen-Zeitung*), which excels through a wealth of favourable location factors. The more straightforward comments on Luxembourg's role as a financial hub include the following examples:

> [...] le pays est devenu en l'espace d'une vingtaine d'années la première place en Europe pour la domiciliation de fonds communs de placement. (*La Tribune*)

> Luxemburg stellt sich als ein Zentrum des Finanz-Know-how vor, das unermüdlich neue und innovative Produkte generiert. (*Börsen-Zeitung*)

The more sophisticated articles from the financial press argue that Luxembourg's success as a centre for private wealth management and particularly for the domiciling and administering of investment funds is based on a whole catalogue of location factors. The most frequently mentioned factors are the adaptable regulatory framework, social and political stability, highly skilled and multilingual staff in the sector itself as well as in the supervisory authorities in charge, agglomeration of specialised business services (e.g. public accountants, and lawyers), good technical infrastructure and favourable geographic location in northwest Europe. The following quotes provide some evidence for the argumentative importance of these location factors:

> [...] l'industrie financière luxembourgeoise et sa croissance reposent sur trois piliers: le cadre réglementaire qui a encore fait l'objet d'aménagements cette année, le volontarisme industriel de l'ensemble de ses acteurs et la protection des investisseurs surveillée par la CSSF. (*La Tribune*)

> These factors, together with laws on holding companies and Luxembourg's political and economic stability, have attracted banks and investors from all over the globe. Good communications, easy access to other European financial centres, skilled multilingual staff, and a tradition of banking professionalism have also contributed to the growth of the financial sector. (*South China Morning Post*)

This wealth of location factors, it is argued, has resulted in the development of a major financial hub that distinguishes itself by its high degree of innovativeness, creativity and flexibility. In particular, the Frankfurt-based financial daily *Börsen-Zeitung* is full of praise for Luxembourg's respective competences:

> Zum internationalen Private Banking und Wealth Management haben sich in Luxemburg über Jahrzehnte hinweg eine ausgezeichnete Kompetenz, Kreativität und ein lösungsorientiertes Handeln etabliert. (*Börsen-Zeitung*)

Overall, Luxembourg's qualities as a financial centre are represented in an extremely positive and assertive way. According to the international specialist press, in this area the Grand Duchy can claim to be a truly metropolitan place where diverse financial networks converge, highly specialised know-how is built, creativity flourishes and innovative power is abundant. Here the micro-state has managed to develop an international appeal that has placed it in the spotlight of the global financial world, or as the *Financial Times Deutschland* puts it: "Im Zeitalter der Informationstechnologien und globalen Finanzströme kommt der Rest der Welt einfach nach Luxemburg."

Dominant framing 6: Transborder integrator par excellence

Compared with Luxembourg's stereotypical image as a banking and finance centre, the scope of transborder integration in the Greater Region is far less present in the public discourse. It does, however, feature in most of the portray-type articles analysed. Given the fact that, for the first time in history, Luxembourg expanded the European Capital of Culture 2007 beyond the confines of one single city (or country) to include the entire Greater Region, and that this event was the prime occasion for many articles to be published, this focus is rather unsurprising.

The main thrust of the text passages on this theme is, yet again, one of positive admiration. The Grand Duchy is, because of its attractive labour market, framed as the main engine of a cross-border integration that is without equal in the European Union:

> Die Regionalität hat in und um Luxemburg herum eine einzigartige Ausprägung erfahren – durch Luxemburg. (*Süddeutsche Zeitung*)
>
> Diese alte Mitte Europas, Kern der Montanunion, spielt erneut die Rolle des Laboratoriums für die Zukunft. (*Die Welt*)

Another predominant discursive thread in this context is the normality of the act of border-crossing, by which the people in this region manage their daily life in economic, social and cultural terms. In this respect, the Greater Region is sometimes seen to be the laboratory of European integration.

> This corner of Europe has always lived with shifting borders; it isn't unusual for individuals to work in one country, only to return to a home in another, while one's mobile phone might be networked into yet one more. Some towns and cities along the Franco/German border feature roads where one side is in France, the other in Germany. Even before the Schengen agreement – named after the nearby Luxembourg town which allows continental Europeans free passage between borders – commutes like this have been a fact of life for people living and working in the area. (*The Guardian*)

> Ganztägig sind in der Großregion Menschen mit luxemburgischen, belgischen, französischen und deutschen Autokennzeichen unterwegs, kaufen bei Aldi im Saarland, beim Kaufhof in Trier, bei Ikea in Belgien, bei Carrefour in Frankreich, und in Luxemburg wird natürlich getankt. Wer dem Strom ein paar Tage folgt und immer wieder die stillgelegten Grenzkontrollhäuschen passiert, erlebt sich bald als Getriebenen eines neuen nachbarschaftlichen Nomadentums, das längst auch politischen Niederschlag abgeworfen hat. (*Süddeutsche Zeitung*)

Although problems arising from this everyday cross-border mobility, such as frequent traffic congestion or political cooperation among the regions involved, are barely discussed, Luxembourg is portrayed as the nourishing centre of a transborder region in which the European dream has become reality.

Meta-narrative 1: Micro-state in transition

Across the lines of the prevalent framings analysed above, Luxembourg is implicitly characterised as a state in transition, constantly adapting to external pressure. In this narrative, two transitions stand out: first, the structural change from a largely steel-based economy in demise to a high-performance service economy; and second, the image-related transition from a staidly place of bureaucratic boredom to a lively cultural metropolis. Both of these transitions are praised as being greatly successful, as the following text passages suggest:

> En fait, le Grand-Duché a réussi sa reconversion quand le secteur sidérurgique est entré en crise dans les années 1970. Il a su se tourner rapidement vers des services à forte valeur ajoutée, notamment vers les services financiers qui représentent un cinquième du PIB. (*Le Monde*)

> Auf dem Bockfelsen gelegen, hat sich die gerade mal 80.000 Einwohner zählende Kleinstadt in den vergangenen zehn Jahren vom Kulturnest zur multikulturellen Metropole gewandelt – und sich hierbei dennoch ihren teilweise provinziellen Charme bewahren können. (*Tageszeitung*)

Meta-narrative 2: Strange land of uniqueness and contrast

Another narrative to be identified in the articles tells the story of a distinctive place that sets itself apart from most other countries in the world by a number of unique features. In addition to some of the dominant framings discussed above (e.g. multiculturalism, prosperity), this uniqueness is primarily attributed to traits related to the country's small size, as the following examples show:

> It's the people. They're different, and they know it. They live in a privileged part of the continent, with different rules, different structures and different attitudes. They earn more, pay less tax, and pay less for what they regard as essentials: wine, cheese, cars and perfumes. (*Sunday Telegraph Australia*)

> Luxembourg is breathtakingly beautiful and it comes with a vibrance, confidence and energy that sets it apart from most of its neighbours. (*Sunday Tasmanian*)
>
> Menschen aus 138 Nationen tummeln sich dann auf einer Fläche, kaum größer als Delmenhorst oder Tübingen. Das ist einsamer Rekord in Europa, vielleicht in der Welt. (*Frankfurter Rundschau*)

Common clichés are deconstructed and modified by elaborating on some of the country's unexpected qualities. The articles, thus, make a contribution towards decoding the "mystery" of the Grand Duchy by throwing light on some of its hidden attributes:

> De Luxembourg, on a l'image d'une place financière et européenne sans beaucoup d'âme et un peu perdue dans les brouillards de l'hiver. Un week-end passé dans la capitale du Grand-Duché infirme en quelques secondes cette image convenue. (*Le Figaro*)
>
> To most of us it's just a place where rich people stash their money, whose part-time footballers occasionally embarrass England's overpaid superstars. However, there's actually far more to Luxembourg than amateur football and private banking, and after decades in the cultural doldrums, it's finally becoming a lively centre for the arts. (*The Guardian*)

Frequently, this narrative of surprise is supported by emphasising the coexistence of well-known, stereotypical features on the one hand and the new and unexpected on the other – the latter being depicted as a well-appreciated makeover of a country formerly known as "a fuddy-duddy place full of bankers and EU bureaucrats" (*The Guardian*). In many cases, this contrast is rhetorically constructed upon pairs of opposites, for instance "provincial" and "cosmopolitan":

> Despite its modest size, this miniature metropolis is remarkably cosmopolitan. (*The Guardian*)
>
> Tiny nation rich in sophistication. (*Sunday Tasmanian*)
>
> Die Weltprovinz. (*Süddeutsche Zeitung*)
>
> [...] hinterwäldlerischer Weltgewandtheit. (*Süddeutsche Zeitung*)
>
> Pulsierende Internationalität und beschauliche Selbstgenügsamkeit begegnen sich auf engstem Raum. (*Tageszeitung*)

Conclusion

In the vast majority of the press articles examined, Luxembourg is represented in a broadly positive light. In general, the texts formed a well-informed impression of the country's recent social and economic history and the associated challenges of change. Given the journalistic nature of the pieces, the coverage on some aspects (e.g. multilingualism), however, lacks accuracy and critical scrutiny. This might partly be based on the aforementioned routines of some journalists/media to uncritically overtake and reproduce clichés or euphemistic information from official (publicity) documents. Nevertheless, hardly any article falls into the trap of inconsiderately reproducing common stereotypes of the Grand Duchy. On the contrary, many articles explicitly aim to deconstruct some of these prejudices by portraying a country that has more to offer than banks and cheap petrol stations, thereby drawing a realistic picture of the country's recent developments and achievements. The Cologne regional daily *Kölner Stadt-Anzeiger*, for instance, straightforwardly titled one of its articles on Luxembourg "Mehr als Banken und Tanken".

This differentiated and predominantly positive picture is confirmed by the vast majority of interviewees. In particular, the recent efforts to make Luxembourg's cultural life more "vibrant" and internationally competitive are unanimously recognised and welcomed. Since Luxembourg's metropolitan image relies heavily on its standing as a major financial centre, the latter's sustainability seems to be less questioned than one could expect. By ascribing it a certain localised know-how both in terms of product development and diffusion in the financial sector as well as in terms of the reactivity and flexibility of the national authorities, most respondents reject the merely simplifying "off-shore" label of Luxembourg as a financial place.

Most of the results presented in this chapter can be found in both empirical sources. This is particularly true for the evaluation of Luxembourg as a business location. Despite this obvious overlap of the data sets, a series of more contradictory or at least more differentiated views could be identified. These concern, in particular, the issue of multiculturalism and social integration on the one hand and cultural life and leisure purposes on the other.

Even though the newspapers almost unanimously highlight Luxembourg's multicultural society as a success story of "harmony" and "good integration", some of the interviewees more critically comment on the inherent risk of the current situation. Describing the various communities as coexisting without overly interacting, the society is framed as being split in two: the Luxembourgish community and the international community, both of which keep themselves to themselves, thereby making it hard for expatriates to get in touch with locals and integrate into their existing networks/circles of friends. Given that the majority of the interviewees were "expats" and part of a wealthy group of foreigners, these findings cannot simply be generalised for other immigrant groups. However, it indicates the potential problems reflected by current debates on Luxembourg's education system and language policies.

The second realm where contradictory statements are clear is Luxembourg's recent development in terms of cultural life. Although almost all newspaper articles try to show that – contrary to prevailing stereotypes – Luxembourg has developed a competitive infrastructure of museums, concert halls and interesting nightlife, this perspective is only partly shared by most interviewees. Even though they acknowledge the recent efforts to install new facilities and establish new cultural events, criticisms of the international attractiveness of Luxembourg almost exclusively target the cultural sector (and, to a lesser extent, the real estate market). The "external" view of journalists visiting Luxembourg from abroad might be more biased by the "surprise" factor of what they found compared with the "insider" view of expatriates experiencing everyday life.

The exploratory nature of this study calls for a more in-depth analysis of the various discourse strands identified. In particular, the few critical voices concerning the resilience of the financial centre, attractiveness of the city and its cultural life and future of Luxembourg's multiculturalism and multilingualism deserve further enquiry. Future research should also take into account how the media reflect recent development trends such as:

- The current debate on "tax paradises" in Europe and how it affects the aforementioned images and evaluations.
- The more general discussion on the Europeanisation of everyday life and to what extent it biases the perception of Luxembourg's multicultural setting.
- The increasing use of terms comprising the prefix "metro" (metropolisation/metropolisation, metropolis, etc.) in the (regional) media or in the political realm with regards to Luxembourg City's economic and cultural standing in the Greater Region, and its potential overrating.

Furthermore, it might be promising to conduct more sophisticated discourse analyses to tap into the question of how and to what extent the media form specific interdiscourse patterns and how the latter influence the attitude of people working in or with Luxembourg.

Part 2

Cross-border Interdependencies

CHAPTER 4

Mapping the Structuring of a Cross-border Metropolis

The Functional Urban Region of Luxembourg

Antoine DECOVILLE and Christophe SOHN

While certain authors expect the 21st century to see the end of the city (Fischman 1998), others expect it to see their ultimate establishment (Carillo 2006), or the "age of generalised urbanisation" (Castells 1999: 295) Whichever view is taken, these predictions do no more than reflect the magnitude of the changes underway in the sphere of the urban fabric, and the difficulty facing researchers attempting to characterise them. It is, however, clear that while the classical characteristics of the city of yesterday may have disappeared, the major urban centres, far from being in decline, appear to be reinforcing their position as centres of influence and drive (Le Galès 2003).

At the root of these transformations are complex economic, technological, political and sociological processes, which can be understood at different levels. The main urban regions are today the engines of the global economy and the areas of integration they create are tending to increasingly put into question the central position of the nation state within the organisation of the global economy (Ohmae 1995, Scott 1998). This urban concentration at the global level is accompanied by a reshaping of the process of urbanisation at the local and regional levels. Hypermobility, rendered possible by the massive use of the automobile, has thus contributed to transforming urban outlines. The city of yesterday – compact, dense, governed by logics of physical proximity to the centre and easily identifiable – has given way to the urban area of today, which is hybrid, loose and governed by logics of connectivity.

As a contemporary form of the process of urbanisation subject to the effects of globalisation (Ascher 1995), metropolisation has become a key concept that covers the transformations of urban forms and scales. Increasingly freed from the constraints of distance with regard to flows of information, capital, commands or knowledge, cities nonetheless remain dependent on an employment area that is indeed located within a local area. It would thus be simplistic to consider urban centres as nodes of globalised networks freed of all local or regional anchorages. The metropolitan centre remains dependent on an area with which it forms a "metropolis-based region" (Bar-El and Parr 2003). Although the metropolitan area may often produce relatively little wealth outside its core, and numerous peri-urbanised towns and villages may be assimilated as no more than dormitory towns, the dependence of the centre of the surrounding area remains, however, real. This zone, which corresponds to the employment area of the metropolis, conditions, to a certain extent, the development capacity of the urban centre.

Focusing on the case of Luxembourg and its peripheral areas within the Greater Region, this chapter aims to examine the form of the functional urban region created through the emergence of a process of cross-border metropolisation focused on the capital city. In order to highlight how the functional interdependencies are shaping the spatial configuration of the cross-border urban region, one has to go beyond static representations and take into account the dynamics at play. There may be multiple factors determining these dynamics, and the aim of this article is to contribute to an understanding of these. In addition, the analysis of the spatial structuring of Luxembourg's metropolitan region will allow us to question the social and territorial cohesion of this space under construction and thus its sustainability.

In the first part, an examination of the concept of the functional urban region allows a definition of the latter to be developed, as well as the approach to be used for its identification. Particular attention is paid to the cross-border context and its methodological implications. The second part focuses on understanding the functional urban region of the Luxembourg agglomeration by means of comparing two complementary cartographical representations. Because this involves a dynamic process and not a fixed configuration, the development of this functional space is then considered in a diachronic manner over a period of fourteen years (1994-2008). Finally, the fourth part examines the socio-residential changes occurring, in particular concerning the redistribution of metropolitan workers and the logics of spatial specialisation.

Functional urban regions: concept and measures

Since the 1950s, multiple works have attempted to analyse the relationship between a city and its environment, giving birth to various concepts including those of the city region, urban area and metropolitan area (see Davoudi 2008 for a critical overview). Following the pioneering work of Berry *et al.* (1968), these areas have generally been conceptualised as functional economic areas. This approach, which is also used in this chapter, focuses on the economic relations defined by the employment areas, with the aid of statistics on home-work commuting. Taking note of these conceptual and methodological choices allows their relevance to the dynamics of cross-border metropolisation at work in Luxembourg to be considered.

Identification of functional urban regions using labour market data

From a conceptual perspective, a functional urban region is characterised by a high intensity of economic and social interactions between a city and various peripheral locations at the intra-regional level. The totality of the spatial entities that are functionally dependent on the urban centre constitutes its area of influence and can be considered as its hinterland. The nature of interactions involved is very wide ranging, as they include commercial exchanges of goods and services, financial transactions, telecommunications and labour flows or trade flows (Brown and Holmes 1971, Karlsson and Olsson 2006).

From an empirical perspective, the identification of functional urban regions can be seen as falling into two major methodological categories: a deductive, top-down approach, which starts from a predetermined collection of urban core areas as the destinations of flows; and an inductive, bottom-up approach, which segments the space in an automatic way on the basis of predefined criteria (Davoudi 2008). In the first type of approach, which is given priority in the present work, the metropolitan core area is distinguished from the commuter attraction area. The latter is generally based on data concerning work-based movements. The intensity of commuting between an urban centre and its peripheral areas is used as an indicator of the degree of economic integration of an employment area. Statistics on commuting have been used in the US since the 1950s to delimit Metropolitan Statistical Areas (Nielsen and Hovgesen 2008). Similar approaches can be found in European countries (Pumain 1993, Davoudi 2008). One of the major advantages of this approach is that it relies on data that is regularly available and that allows the dynamism of the labour market and the area of influence of a city to be taken into account. The residential localisation of the labour force allows the outlines of a functional space to be established, which, although it may appear to be as yet little urbanised in certain places, will tend to develop under the influence of the city centre on which it depends.

The use of home-work commuting data does not, however, allow the full complexity of interactions between an urban centre and its area of influence to be taken into account. First, as emphasised by Davoudi (2008), the relations between a city and its surrounding region are not limited to the daily mobility of workers, or even to flows of economic activity; a multiplicity of social, cultural and environmental interactions is also involved in the structuring of city regions or urban areas. In addition, each of these linkages creates its own spatial imaginary with limits that differ from one to the other. Secondly, given a process of metropolisation that is taking place within a territorial framework inherited from the past, and projecting the modern forms of urban life beyond the city in the strict sense, statistical or cartographic representations encounter several difficulties in dealing with a space under construction for which the geography of internal flows tend to become more complex, especially due to the emergence of "peripheral centres". The delimitation of a functional urban region is rendered difficult, especially by the absence of clear boundaries; instead, what is found is a fuzzy reality structured by gradients with fluid boundaries (Harding *et al.* 2006). Furthermore, the dilution of the urban beyond urban agglomerations no longer obeys the principle of spatial continuity but rather leads to the creation of discontinuous and multipolar functional ensembles. What is at issue are shifting spaces whose internal structures and external limits are evolving as a result of a combination of macroscopic factors and individual decisions made by a multiplicity of actors. In a cross-border context, the complexity inherent to the phenomenon being studied is multiplied by the methodological complications.

Methodological choices constrained by the taking into account of the cross-border context

The cartographic representation of employees at their place of residence poses, in a cross-border context, a methodological problem in relation to the choice of the geographical scale at which the information is reported. Certain authors have already discussed at length this phenomenon, known by the acronym MAUP (Modifiable Areal Unit Problem) and "arising from the imposition of artificial units of spatial reporting on geographical phenomenon resulting in the generation of artificial spatial patterns" (Heywood 1998). This problem, which is a significant issue for studies concerning the flow of home-work mobility (Horner and Murray 2002), is even more pronounced in a cross-border context, with a combination of differing statistical reference units and different local territorial units.

To address this problem, it is necessary to define new statistical reference units using a relevant comparison procedure. While the preferred spatial unit is generally the *commune* (municipality), we have chosen to work at the scale of the *canton* (district) in Lorraine and in Luxembourg, and the *Gemeindeverband* (grouping of municipalities) in Rhineland-Palatinate, in order to have units more homogeneous in terms of population and size, as shown in Table 4.1. These spatial aggregations have allowed the difference between the average population of the different units to be reduced from 1:20 to 1:3 between the two extremes, and the difference in terms of size to be reduced from 1:7.5 to 1:4.3. Such a reduction in the differences between territorial structures improves the comparability of the phenomenon studied between two countries, and facilitates the study of spatial dynamics at the cross-border level.

Table 4.1 – Comparison of spatial units of analysis

	Administrative reference unit	Number of reference units	Average population per reference unit	Average size of reference unit (km^2)
	Data prior to spatial aggregation			
Saarland	Commune	52	20,482	49.4
Wallonia	Commune	262	12,819	64.3
Lorraine	Commune	2,337	1,016	10.1
Rhineland-Palatinate	Commune	2,306	1,756	8.6
Luxembourg	Commune	118	3,725	21.9
	Data after spatial aggregation			
Saarland	Commune	52	20,482	49.4
Wallonia	Commune	262	12,819	64.3
Lorraine	Canton	156	15,214	150.9
Rhineland-Palatinate	Gemeindeverband	191	21,199	103.9
Luxembourg	Canton	12	36,628	215.5

Sources: Eurogeographics, ESPON. Calculations: Decoville and Sohn.

The use of statistics on commuting between places of residence and places of work constitutes another obstacle, due especially to the cross-border nature of the flows between the four countries of the Greater Region. Taking into account national differences in the production of statistics and the problems of comparability that these generate, we have opted to use data from a single source: the General Inspectorate for Social Security (Inspection générale de la sécurité sociale, IGSS). The administrative files produced by the IGSS provide information on all persons covered by Luxembourgish social security on March 31 each year. This therefore includes employees resident in Luxembourg, cross-border employees and all of their co-insured family members. For each of these people, the place of residence, place of work (on the basis of the head office of the company) and the type of professional activity are available. Apart from the presence of this information, one of the advantages of this data is that it is updated annually and thus diachronic analyses from 1994 up to the present are possible.

Cartographic representations of the Luxembourg functional urban area

In order to understand the cross-border metropolitan region emerging around Luxembourg, this chapter takes a deductive approach based on the a priori identification of the urban core which attracts flows of commuters; in this case, we examine the urban agglomeration of Luxembourg, including the capital city and the eight peripheral agglomerated municipalities (Sohn 2006). Taking into account the criticisms previously cited about this type of analysis based on the use of labour market data, it seemed desirable to offer two different interpretations of the functional urban region of the agglomeration of Luxembourg. By changing the way in which the intensity of commuting within the cross-border employment area is represented, one changes significantly the spatial extent of the phenomenon being examined. This approach stipulates that there is not a functional urban region that imposes itself but rather that there are different representations of the same phenomenon. Through the combination of these representations, each of which illustrates a different aspect, one can ultimately come to understand a multifaceted reality.

Representation of proportion of commuters in the resident population: a structure with concentric gradients

The map of the proportion of commuters into the agglomeration of Luxembourg, in relation to the total resident population of the territorial units, offers an initial representation of the area of influence of the Luxembourgish capital for the year 2008 (see Map 4.1). The denominator used is a default choice as there is no data on the population employed at the cross-border level. An initial observation that can be made is that the proportion of persons working in the main agglomeration, within the total population of the territorial units considered, logically falls as a function of the distance from the place of work. The areas most strongly linked to the main urban centre from the point of view of the labour market are located close to the centre, while the areas less influenced by it are at the periphery.

Map 4.1 – Commuter attraction area of the agglomeration of Luxembourg, 2008

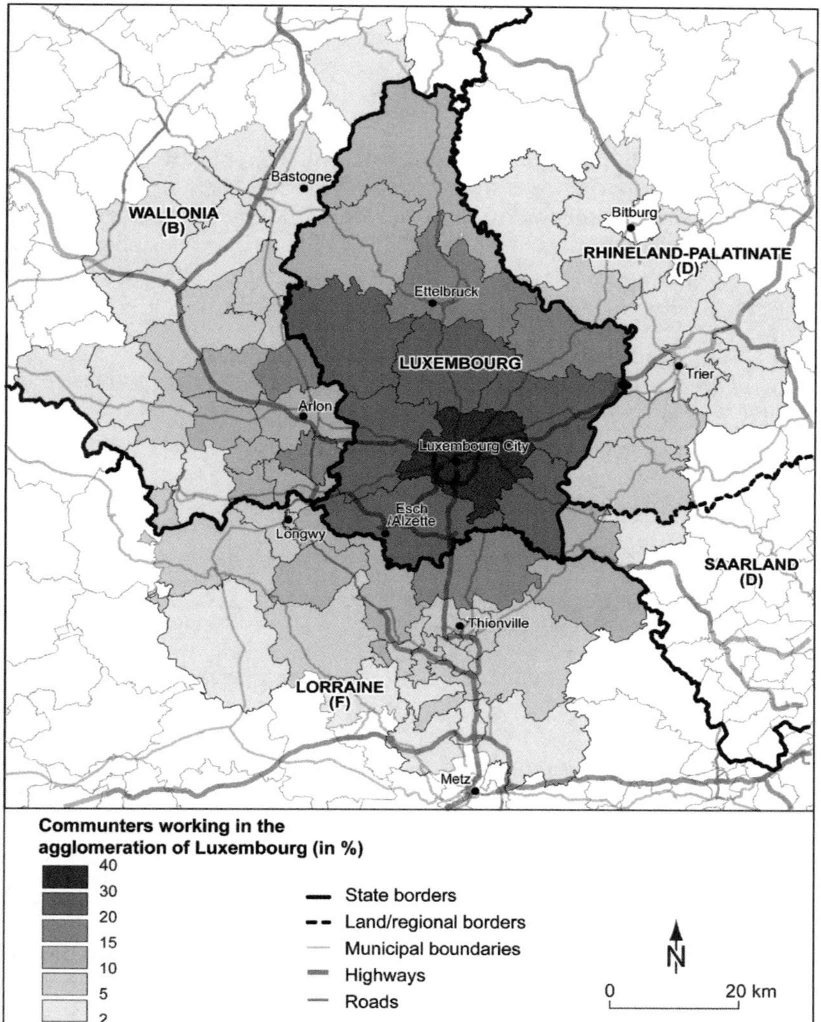

Sources: IGSS 2008.

Authors: Decoville and Sohn, 2009.

When those spatial entities with a commuting rate of at least 10% of their resident population – i.e. approximately 20% of the active population – are considered as part of the employment area of the agglomeration of Luxembourg, the resulting functional region covers the whole of the country and the French and Belgian territorial units situated close to the border or to the major roads (in the case of Belgium). This highly polarised area covers 3,856 km^2 and has almost 660,000 inhabitants, which means it is 50% larger than the size of Luxembourg and has 40% more inhabitants. The more one reduces the minimum proportion of commuters, the more the functional urban region expands and the more inhabitants it has. That said, with two per cent of commuters in the resident population, representing about five per cent of those in employment, the margins of this functional space remain only weakly linked to the phenomenon of cross-border working with the agglomeration of Luxembourg as its destination, with certain peripheral areas being more heavily affected by other urban centres within the Greater Region such as Metz, Saarbrücken and Liège.

Table 4.2 – Characteristics of the functional urban region as a function of the proportion of commuters, 2008

Commuters as proportion of resident population (as % of the active population)	Commuters	Population	Size (in km^2)
At least 20% (~ 40%)	111,549	408,674	1,515
At least 10% (~ 20%)	147,502	659,630	3,856
At least 5% (~ 10%)	171,842	1,002,874	5,611
At least 2% (~ 4%)	184,413	1,393,735	8,888

Sources: IGSS, 2008. Authors' calculations.

The cross-border dimension of the functional urban region of the agglomeration of Luxembourg is the other striking fact revealed by the analysis. Of the 197,000 workers with a job within the agglomeration of the capital of Luxembourg, 72,300 reside within the neighbouring German, Belgian or French territories, representing around 37%. The intensity of the phenomenon within the border regions is, however, highly variable, as at least 10% of the resident populations in certain French *cantons* and Belgian *communes* are commuters, while other border areas have much lower rates.

Two factors need to be distinguished in order to explain these disparities. First, there is the effect of geographical distance, or more precisely the time required to cover a distance, depending on the transportation system. That certain border territories have low rates of commuting is due primarily to the fact that they are relatively far from Luxembourg City; the peripheral ring corresponding to spatial units with between two and five per cent of commuters is thus located on average one hour by road from the Luxembourgish capital.

The other explanatory factor concerns the possible existence of a barrier effect linked to the presence of national boundaries. Within the border strip located approximately 30 km from Luxembourg City, the proportions of commuters are systematically lower than the rates seen in the adjacent Luxembourgish *cantons*, but there is a very clear difference between the German side and the French and Belgian sides (see Table 4.3). In the case of Germany, the spatial discontinuity is especially marked because the difference between two sides of the national divide is one of 1:4, while in the other two cases the difference is smaller (1:2). Indeed, within the Belgian and French areas, certain spatial units that border Luxembourg, such as the Belgian *communes* of Messancy and Attert, and the French *canton* of Cattenom, have values close to those seen in Luxembourg.

Table 4.3 – Proportion of commuters and border effects, 2008

	Country of border regions in question			
Frontier	Luxembourg	France	Belgium	Germany
Luxembourg-France	22.6%	13.2%		
Luxembourg-Belgium	22.7%		14.0%	
Luxembourg-Germany	24.3%			6.2%

Sources: IGSS, 2008. Authors' calculations.

While such a representation, based on the intensity of the phenomena of work commuting within each spatial entity, indicates the degree of economic integration of these territories within an employment area centred around the agglomeration of Luxembourg, it does not give any useful information about the significance of the phenomenon in absolute terms. As an example, the city of Metz, which is located outside of the area of influence of Luxembourg, has over 2,200 people who go to work daily in the Luxembourgish agglomeration, while the *commune* of Messancy, which is seen to be proportionally highly integrated into the Luxembourgish economy, has only 1,291. It is necessary to complement this partial representation of the functional urban region with a different cartography.

Representation by density of commuters: a discontinuous and multipolar space

By choosing to represent the density of commuters working within the agglomeration of Luxembourg, the emphasis is put on the geographical distribution of workers within the functional urban region without seeking to play down the importance of the phenomenon as a function of the resident population of each territory. The representation of densities of commuters per square kilometre, rather than the use of absolute figures, has the advantage of eliminating the effects of the size of the spatial units in question. The cartographic smoothing carried out in Map

4.2 also allows the cartography to be freed from the administrative boundaries to express the phenomenon continuously across space.

This method of representation shows, first of all, the important role of urban centres as providers of workers to the Luxembourgish capital. Apart from the agglomeration of Luxembourg, which is seen as a major residential centre, peripheral urban centres can also be discerned. Thus, the former steel area located across the south of Luxembourg and the French border region that runs from Audun-le-Tiche to Longwy has especially high densities of commuters. In the border regions, the agglomeration of Thionville and, to a lesser extent, those of Metz and Trier, also appear as privileged places of residence of cross-border commuters. The availability of accommodation, and the amenities and services provided explain the residential attractiveness of cities and peripheral urban agglomerations.

Within this geography, the low-density rural zones appear, logically, to be at the back. The Belgian Ardennes, the Eifel national park in Germany and the north of the department of Meuse in France, all peripheral regions that are weakly populated and located away from the major communication routes, have especially low absolute levels of employment within the Luxembourg agglomeration. Likewise, the interstitial spaces that separate the urban centres experience "tunnel effects" along the major roads, especially in the Lorraine corridor between Metz and Thionville.

In contrast to the first cartographical representation, which allowed an assessment of the relative number of commuters in relation to their territorial unit of residence, this second approach emphasises the importance of the phenomenon with respect to the geographical distribution of workers employed within the agglomeration of Luxembourg. The near-regular decreasing gradients from the Luxembourgish centre of employment towards its periphery is replaced by a representation marked by urban centres. After having examined the current configuration of the functional urban region of Luxembourg from multiple angles, we will now take a temporal approach in order to better establish the process at work behind the structure.

Map 4.2 – Density of commuters working in the agglomeration of Luxembourg by place of residence, 2008 (see colour section)

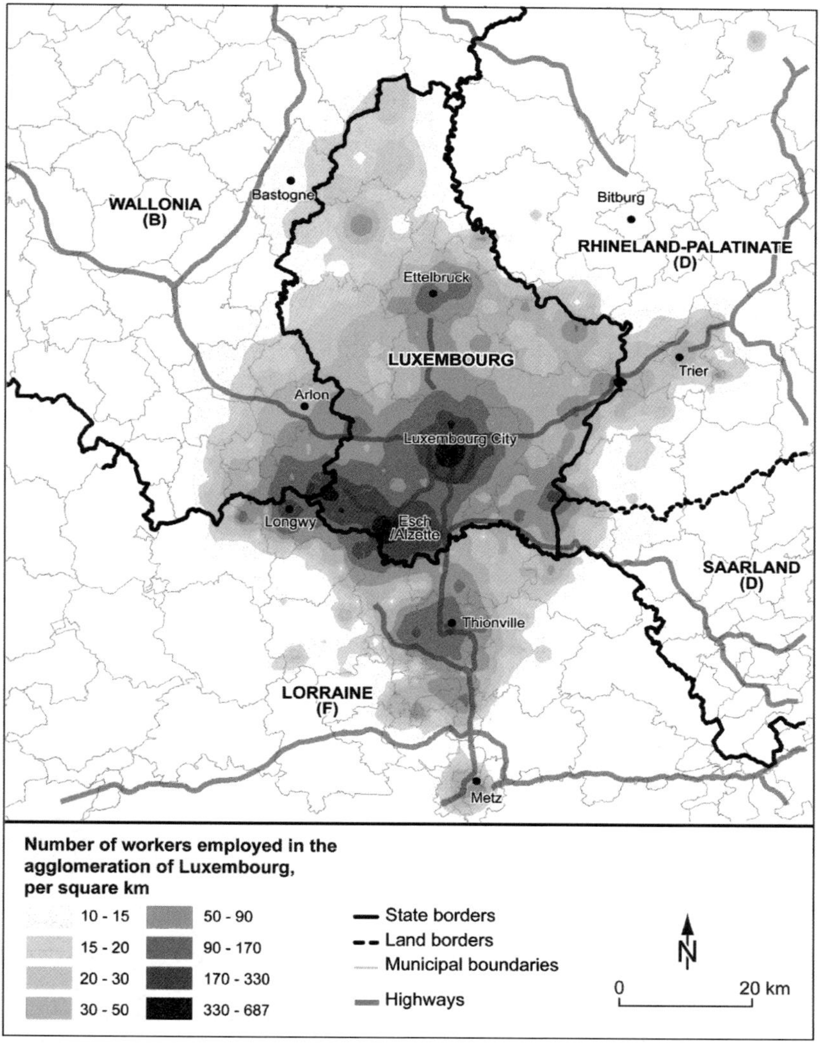

Sources: IGSS 2008.

Authors: Decoville and Sohn, 2009.

Dynamics of the functional urban region: between expansion and spatial restructuring

To understand the dynamics that shape the functional urban region over time, two processes need to be distinguished. First, there is the phenomenon of the expansion of the area of influence of the agglomeration of Luxembourg over the course of the progressive integration of new territorial areas. This phenomenon could be an expression of either an urban loosening in terms of residential localisation or a growing polarisation of the territory in terms of employment. Second, the internal process of reorganisation of the functional urban region may operate as an effect of an increasing density of commuters in the high-growth regions and, possibly, decline in poorly accessible areas or those offering an unattractive living situation.

As in the analysis conducted in the first part of this article, two approaches are used to examine the different aspects of the changes that affected the functional urban region of Luxembourg between 1994 and 2008.

Spatial expansion of the area of influence of the agglomeration of Luxembourg: an uneven development

The map of gradients showing the proportions of commuters of two and 10% in 1994, 2002 and 2008 shows the expansion of the area of influence of the Luxembourgish agglomeration (Map 4.3). For the 10% threshold, which represents areas strongly integrated into the economy of Luxembourg City, we move from a situation in 1994 where, with the exception of one Belgian *commune*, the whole of the area of influence is located within Luxembourg to a situation in 2008 where almost one third of the area is located within the border regions, primarily in France and Belgium. It can be seen from Table 4.4 that it is above all between 1994 and 2002 that the spatial expansion of the 10% area of influence was at its strongest (+29.1%), with growth having slowed more recently (+12.7%). When the development of the area of influence is examined using the less restrictive threshold (two per cent commuters) in order to assess the spatial dynamic at its margins, a similar tendency can be seen, with strong growth between 1994 and 2002 (+52%), especially in Germany and Belgium, followed by more limited growth between 2002 and 2008 (+19.8%).

Although comparison between different gradients is not easy, these results corroborate the widely held idea that the further one moves from the urban centre the harder integration within its area of influence is, given the way in which distance acts as a brake. The dynamic of spatial expansion of the functional urban region of Luxembourg thus tends to slow as it expands. That said, this geographical expansion in no way prejudices the contribution of new areas integrated into the employment area in terms of the number of additional workers.

Map 4.3 – Spatial expansion of the area of influence of the agglomeration of Luxembourg, 1994-2008 (see colour section)

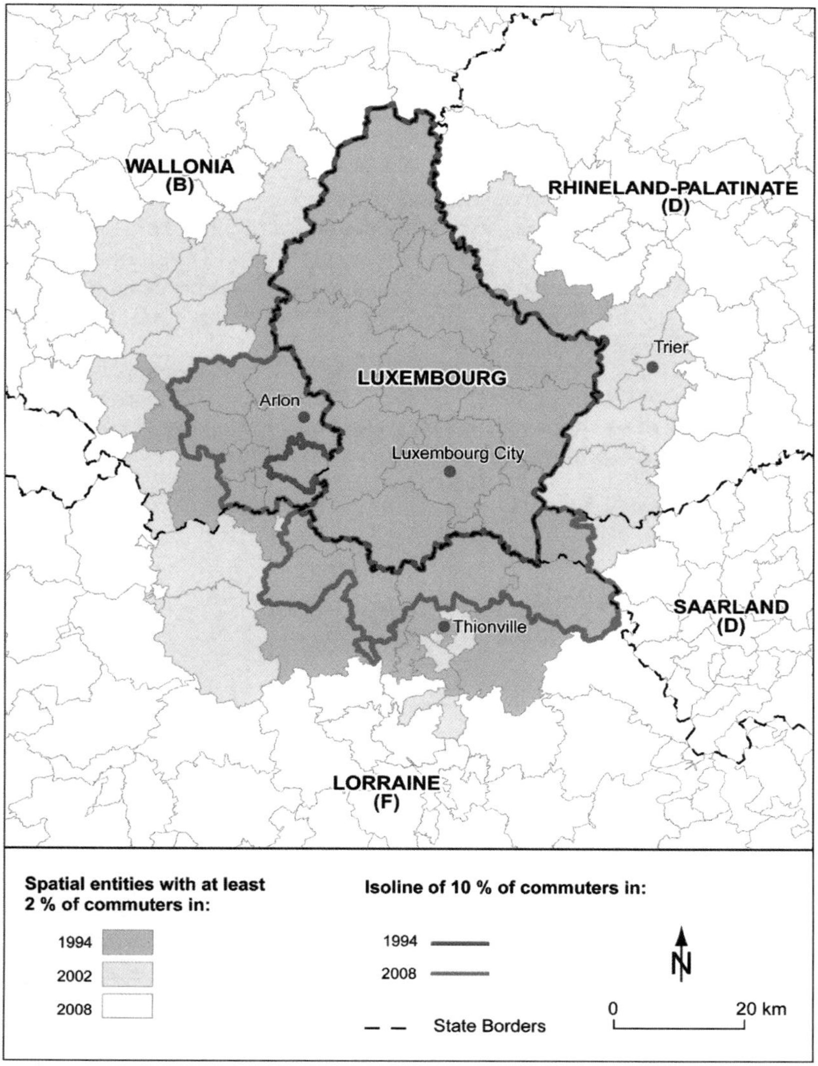

Sources: IGSS 1994, 2002, 2008.

Authors: Decoville and Sohn, 2009.

Table 4.4 – Spatial expansion of the area of influence, 1994-2008

Proportion of commuters	1994	2002	2008	1994-2002 (%)	2002-2008 (%)
2%	4,879	7,416	8,888	2,537 (+52)	1,472 (+19.8)
10%	2,649	3,421	3,856	772 (+29.1)	435 (+12.7)

Sources: IGSS, 1994, 2002, 2008. Authors' calculations.

To better evaluate the contribution represented by the successive expansions of the area of influence of the agglomeration of Luxembourg over time, it is possible to measure both their absolute and relative contributions in terms of number of workers (Figure 4.1). The identification of the territories constituting the area of influence in 2008 is based on Map 4.3 and uses the two per cent commuters threshold to take into account the majority of flows of workers into the agglomeration of Luxembourg and to thus provide a broad view of the process. It should be recalled that this is a lower threshold and that many communes have a much higher rate of commuting.

Figure 4.1 – Change in the number of commuters as a function of territories constituting the area of influence of the agglomeration of Luxembourg

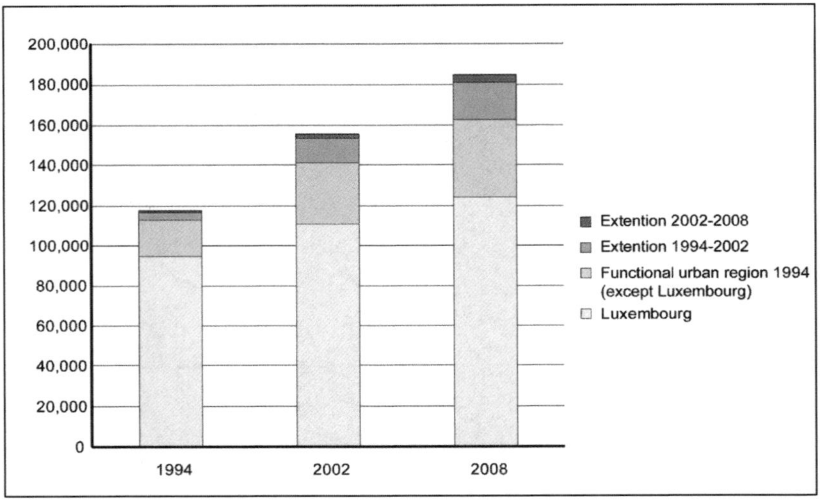

Sources: IGSS, 1994, 2002, 2008. Authors' calculations.

Note: 2% threshold for commuters.

In absolute terms, not only does Luxembourg represent the main provider of workers to its main urban agglomeration on the different dates shown, but it is also within the country that the growth of people in work is highest. The contribution of the successive extensions to the area of influence at the level of the border regions decreases as a function of the period of integration: the territories integrated most recently into the area of influence of the agglomeration of Luxembourg tend to make a weaker contribution than those that have been integrated for longer.

In order to clarify this dynamic of restructuring, a detailed cartography of the development of the relative contribution of the areas to the growth of employment within the agglomeration is necessary.

Relative contribution of areas: towards a redistribution to the border areas

The analysis of the distribution of commuters within the functional urban region of Luxembourg and its changes over time is based on a relative approach (in relation to the total employment within the Luxembourg agglomeration). A fall in the relative contribution of a territorial unit does not therefore necessarily correspond to a fall in the number of persons in employment. As all territories integrated into the functional region saw an increase in the number of their residents employed within the Luxembourg agglomeration between these two dates, the variations shown in Map 4.4 express levels of growth above or below the average.

In general, this cartographic representation clearly shows the fall in the relative importance of Luxembourg as a place of residence, with a corresponding increase in the importance of the peripheral border areas. This development is an expression of a phenomenon of internal spatial differentiation that benefits, from a residential point of view, the border urban areas, while employment continues to grow within the metropolitan core. Thus, the urban agglomerations of Trier, Thionville and Metz are the areas that have seen the strongest growth in terms of their contribution to employment within the Luxembourg agglomeration. In the rural regions, such as the Belgian Ardennes, the relative growth is lower. Finally, in Luxembourg, only the communes of Frisange and Strassen have seen their number of residents employed within the agglomeration grow more strongly and relatively than the average for the communes within the area of influence.

Among the areas that have seen their relative weight fall, one can point first of all to the agglomeration of Luxembourg and secondly to the Franco-Luxembourgish steel area based around the city of Esch-sur-Alzette. For Luxembourg City and its periphery, the proportion accounted for by its residents within the employment area of the agglomeration fell from almost 30.8% in 1994 to 23.7% in 2008, despite an increase of almost 10,000 commuters within the area.

Map 4.4 – Relative changes to density of commuters to the Luxembourg agglomeration, 1994-2008 (see colour section)

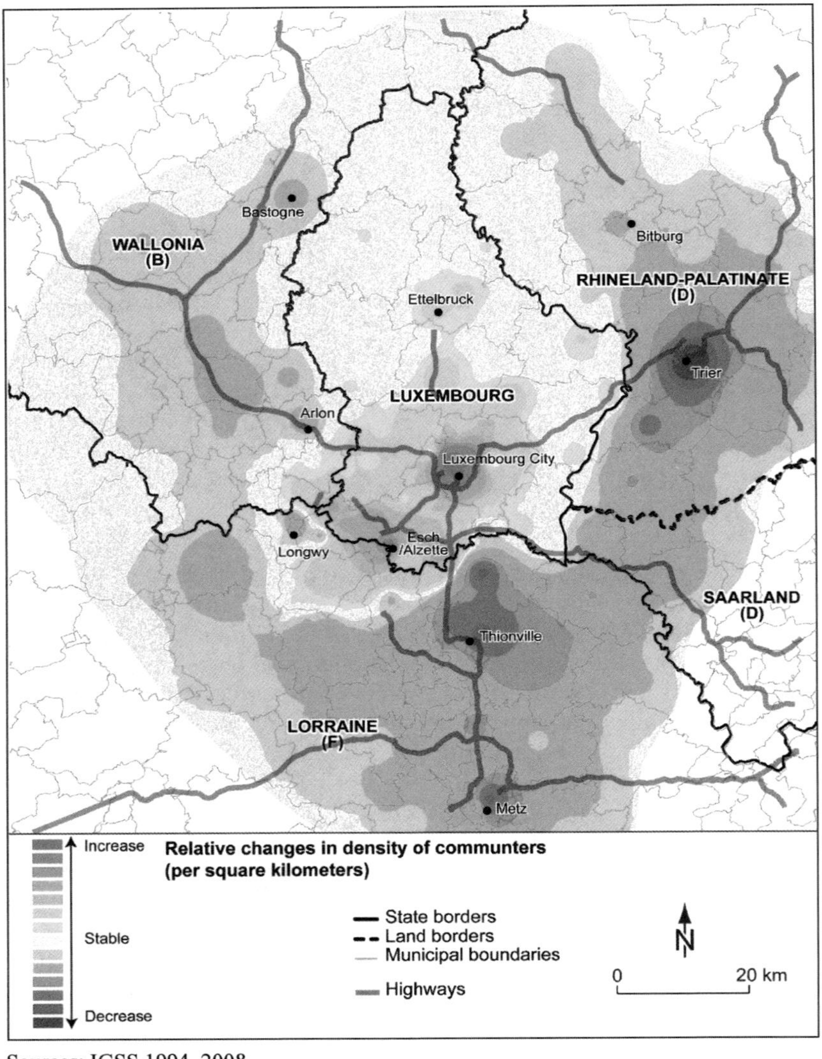

Sources: IGSS 1994, 2008.

Authors: Decoville and Sohn, 2009.

This data shows the increasing cross-border character of the functional urban area of Luxembourg City, and highlights the processes of the strengthening or weakening of the different areas that it is made up of. However, they do not allow the residential logics of the workers to be distinguished; these are heavily conditioned by the type of employment.

Towards a specialisation of areas within the metropolitan region?

In this third and final part of the analysis, the socio-residential structure of the functional urban region of Luxembourg will be examined by means of the type of employment of commuters, with metropolisation being understood as involving both a concentration of knowledge-intensive service activities and of technology-intensive industrial activities (Krätke 2007). Complementing the analysis conducted by Walther (Chapter 1) concerning the localisation of these activities within the metropolitan area, here consideration will be given to the place of residence of the commuters, distinguishing between commuters who have employment strongly linked to the knowledge economy and the others.

Taking as a base the nomenclature of "metropolitan employment", defined by Sohn and Walther (2009) for Luxembourg, and inspired by the work of Eurostat (2006), it is possible to distinguish between areas in which a large proportion of those employed within the Luxembourg agglomeration work in fields related to the knowledge economy which drives economic growth within the metropolitan centre, and areas which are clearly less "metropolitan", in which those working in the metropolis tend to be less qualified. This analysis ultimately allows questions relating to the social and territorial cohesion of the functional urban region to be addressed.

Proportion of metropolitan employment: an "archipelago structure"

As shown in Map 4.5, the highest proportions of commuters with a knowledge-intensive job live in the urban centres and in the interstitial areas along the main roads. Conversely, non-metropolitan employment is found predominantly in the rural regions, such as the countryside of Longwy (France), the region of the Südeifel national park (Germany) and the northern part of Luxembourg (Oesling region).

Map 4.5 – Place of residence of metropolitan commuters, 2008 (see colour section)

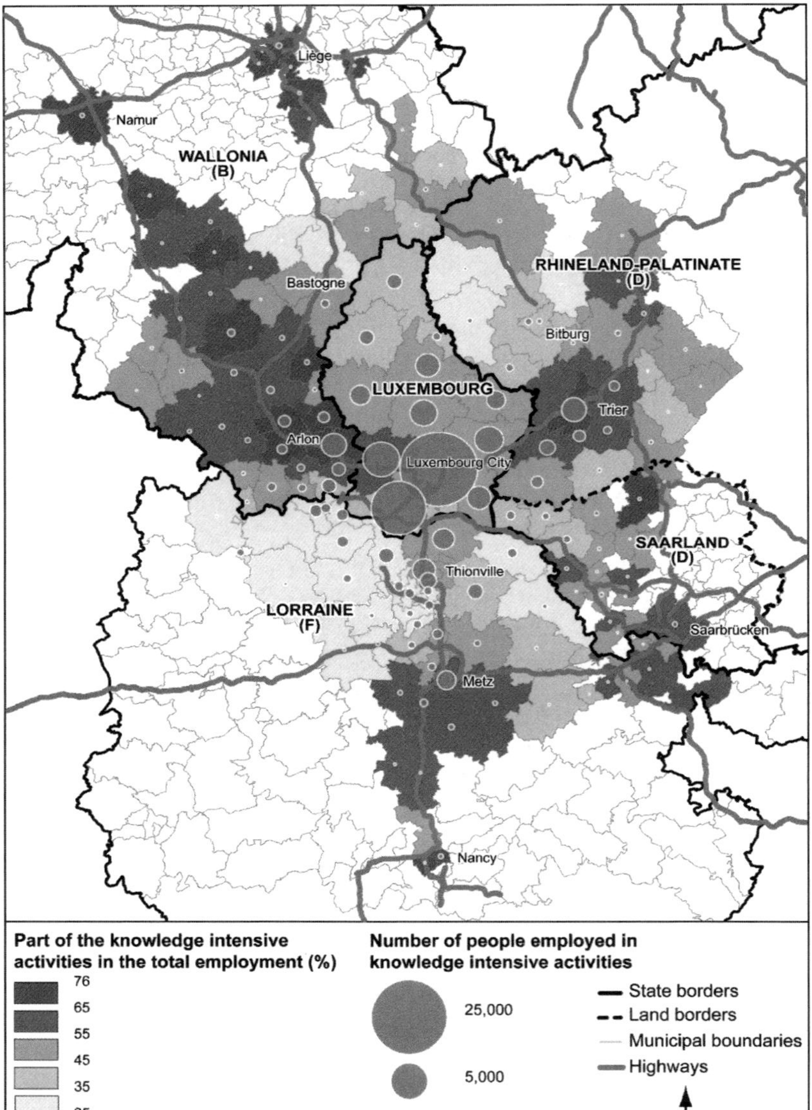

Sources: IGSS 2008.

Authors: Decoville and Sohn, 2009.

The distance from the place of work also appears to be a criterion that distinguishes the places of residence of commuters working in knowledge-intensive sectors from others. The further employees in urban centres live away from the agglomeration of Luxembourg, the higher the proportion of them working in knowledge-intensive sectors. The three territorial units with the highest proportion of knowledge-intensive employment within the total employment of the Luxembourg agglomeration are three medium-sized towns located over 100 km from the centre of employment: Liège, with 76%; Namur, with 74.5%; and Nancy, with 71.6%. On the other hand, the territorial units with the lowest relative proportions of knowledge-intensive employment are located in rural areas, relatively close to the Luxembourg border, or in the former steel areas.

This data supports the idea that highly-qualified individuals prefer an urban lifestyle, while less qualified workers appear to place more importance on the balance between distance to work and property prices. In addition, those commuters with a job in a knowledge-intensive sector (finance, business services etc.) are better able to bear the additional transport costs resulting from the greater distance from their place of residence, due to their attractive rates of pay. Also, the spatial configuration of the functional urban region has a different profile depending on the type of employment examined. For knowledge-intensive activities, the area of influence of the Luxembourg agglomeration is larger, with an archipelago structure arranged around the urban centres.

Average distance between home and work: socio-economic disparities

In order to understand the spatial distribution of the places of residence of commuters who travel daily to the agglomeration of Luxembourg, and to see the disparities according to their socio-professional profile (metropolitan and non-metropolitan jobs), it is useful to calculate the standard distance (see Table 4.5). This Euclidian value represents the average distance between the places of residence and of work. Taking into account the geographical location of Luxembourg City in relation to the border regions, the analysis distinguishes between commuters resident in Luxembourg and those who live across a border, the majority of the former living closer to their place of work than do the latter (see Map 4.5).

Table 4.5 – Average distance between home and work of metropolitan and non-metropolitan commuters (in km)

	Residents		**Cross-border commuters**	
	Metropolitan employment	Non-metropolitan employment	Metropolitan employment	Non-metropolitan employment
1994	12.3	11.8	37.7	32.4
2008	11.9	13.3	42.0	37.7
Change (%)	-3.3	+12.7	+11.4	+16.4

Sources: IGSS, 1994, 2008. Authors' calculations.

For residents of Luxembourg, one sees a slight spatial concentration of metropolitan workers and a relative dispersion of non-metropolitan workers. One can see in this contrasting trend the effects of the real estate pressure that applies to the Luxembourg agglomeration and its peri-urban surrounding areas; the increased property prices tend to push less-qualified workers to the periphery.

For the cross-border commuters, the trend is towards the spatial dispersion of places of residence, both for those with a metropolitan job and for the others. That said, in contrast to the resident commuters the metropolitan workers live on average further from their place of work than the non-metropolitan workers, although the difference between the two tended to fall between 1994 and 2002. This observation may be surprising at first glance, as the metropolitan jobs are on average better paid and these workers thus have, potentially, access to a home closer to the centre. This would be, however, to underestimate the multipolarity of the area in question and the great diversity of the factors that can be involved in the choice of place of residence of the highly qualified workers with jobs within the Luxembourg agglomeration. As a function of the origin of the highly qualified workers, their attachment to their place of origin, and their desire to enjoy the services and amenities offered by an urban location, the logic of proximity to the centre becomes less important (see Gerber, Klein, and Carpentier, Chapter 6). For those with the resources that allow them to make a real choice about their place of residence, it appears that the urban lifestyle is preferred (van Winden, van den Berg, and Pol 2007). The importance of metropolitan employment in the urban centres distant from the Greater Region (Nancy, Namur, Liège, Saarbrücken) appears to confirm this statement.

Conclusion

As a mental construction, all representations tend to privilege certain criteria and neglect others. A map is thus never neutral but rather conveys a series of representations, which are often ideologically coloured. The cartographic representations analysed in this chapter are no exception to this rule; whether the focus is placed on the proportion of commuters as a function of the resident population or on the density of commuters per square kilometre, each map is based on choices, motivated by methodological consideration but also reflecting a certain spatial imaginary of the functional urban region of Luxembourg. We have thus been able to see that in the first case the effects of distance are privileged, which shows concentric gradients focused on the capital, while in the second representation a discontinuous and multipolar space is shown.

In light of this, it would appear illusory to seek to find a unique and universally valid definition of the extent of the functional urban region of Luxembourg. It appears more constructive and interesting to attempt to understand the nature of the processes at work and the factors determining them. In concrete terms, this study of the area of influence of the Luxembourgish urban pole has provided clarification of the dynamics of the expansion and residential specialisation within the transnational employment area. It is thus clear that the functional urban region of Luxembourg is tending to become, proportionally, increasingly cross-border in nature. However, the dynamic of the territorial expansion of this area appears today to be running out of steam, due to the brake exercised by distance. The process of metropolisation operates today more by a reinforcement of the attraction exercised by the agglomeration over the municipalities already integrated into the metropolitan area, with the existence of elements driving the integration process, such as communication pathways. Finally, it is important to note that this functional area is subject to residential logics that differ as a function of the sectors within the workers employed in the Luxembourg agglomeration work. As is the case for other metropolitan regions in Europe, the average distance travelled by highly qualified commuters is greater than for non-qualified workers (see in particular Coombes and Wymer 2001). As well as the preference of individual workers in the knowledge-intensive sectors for urban locations, this phenomenon underlines that the spatial expansion of the employment area varies as a function of the socio-professional categories examined. Firmly shifting and multifaceted, the functional urban region of Luxembourg remains characterised by fluid and tangled boundaries.

CHAPTER 5

Local Mobility and Cross-border Peri-urbanisation

Philippe GERBER, Olivier KLEIN and Samuel CARPENTIER

Due to its cross-border characteristics, the specific features of the Luxembourg metropolitan area have given rise to unique functional specialisation. A classic centre-periphery polarisation is combined with cross-border differentials particularly in terms of wages (Berger 2005) and real estate prices. This reflects the paradoxical role that European state borders are likely to occupy. Briefly, this role can be divided into two: interface (stimulation, interaction) and barrier (filter, separation), (Ratti 1994). On the one hand, the interface role is illustrated by the implementation of the European Union, (*e.g.* border opening, single currency). Then, the "birth" of the cross-border worker as a legal entity, recognised as such in 1971 through the EEC Regulation No. 1408/71 (4), or the Schengen Agreement of 1985, are only a few illustrations facilitating both the search for employment outside the country of residence and cross-border residential moves. On the other hand, the role of barrier reflects the existence of cross-border differentials such as wages or real estate prices, but also regarding social relations (gender, family, etc.), territorial attachment or identity issues (language, etc.), (Legrand 2002).

The analysis of the local mobility system, combining daily movements and residential mobility within this cross-border metropolitan area, allows a better understanding of the details and spatial consequences of this functional specialisation and its related border effect. Addressing both journeys to work practices and the residential choices of cross-border and resident workers, this chapter illustrates the functional interdependencies that exist on either side of the Luxembourgish border. In the first section, we examine the concept of peri-urbanisation and its links to the process of metropolisation. Then, a first analysis of the journeys to work flows and the residential journeys of Luxembourg's workers will show the current patterns taken by this peri-urbanisation. Finally, using cartography of vector fields, inspired by the work of Tobler (1976), the role of the border within the system of local mobility will be demonstrated.

Local system of mobility and peri-urbanisation

Even if metropolisation has its roots in global flows, crystallised in city networks (Castells 1989), it nonetheless has numerous local implications, including peri-urbanisation.

Metropolisation and peri-urbanisation

Residential sprawl, which has now been shaping contemporary landscapes for over a century, can be summarised as having two major chronological stages. The first is one of morphological sprawl, linked in particular to economic (the industrial revolution) and demographic growth (the rural exodus and the increase in the population). These concomitant developments, accompanied by the development of public transport, first supported the development of inner suburbs adjoining the city centres in the 19th and at the start of the 20th century (Pinol 1992), and then of suburbs taking the form either of detached houses or collective housing (Howard 1985). The second stage is more recent; this concerns essentially a functional spread of residence, which is no longer contiguous with the city centre (Dezert *et al.* 1991) but which remains easily accessible thanks to individual modes of transportation such as the car. This phenomenon is linked to the "Thirty Glorious Years" between 1945 and 1975, characterised by a large increase in the standard of living, the baby boom, a high level of social mobility (Fourastié 1989) and undeniable progress in terms of accessibility (Wiel 1999). The functional spread was also reinforced by the process of metropolisation. This second stage ultimately included the successive stages of peri-urbanisation, identified from the end of the 1960s in the United States by Webber (1964) and in France by Bauer and Roux (1976).

Whatever is understood by the terms peri-urbanisation, suburbanisation or counter-urbanisation (see Bauer and Roux 1976, Jackson 1985, Berry 1976), it is clear that, since that time these peripheral areas of major agglomerations have experienced the majority of metropolitan demographic growth in the western world, even if intra-urban regeneration such as gentrification (Gerber 2000) is being seen in the developed countries.

In all cases, peri-urban development has now overtaken the suburbanisation of the industrial period to cover a large part of the rural world. Independently of territorial administrative borders, the metropolises articulate, on the one hand, flows on the global level, thus constituting a veritable "archipelago economy" (Veltz 1996) and, on the other hand, the mobility of goods and persons at the local level (Orfeuil 2000). It is on this latter aspect that this chapter will focus.

The local system of mobility

The development of functional residential sprawl, at least since the end of the 20th century, is the result of increasing means and capacities of daily transportation, in particular as a result of the democratisation of the use of the automobile (Dupuy 1999). This latter development allowed a large majority of the working population to expand their possibilities in terms of residential mobility while keeping the same work place (Brun and Bonvalet 2002). In addition, the real estate market has become significantly tougher, with ever-increasing costs, especially in urban centres, further increasing land rents; as a result, the places where daily activities take place, in particular places of residence and work, are increasingly far from each other (Tabourin *et al.* 1995) and daily life is more and more fragmented (Le Breton 2005).

Even if it is clear that daily mobility and residential mobility are inextricably linked, because "daily movements are conditioned by the residential choices, which are implied by the migratory options and vice versa" (Bottai and Barsotti 2003), there exist, as far as we are aware, only very few studies that address this. However, Bulhardt and Bassand (1981) referred in the early 1980s to the need to understand spatial mobility as a system, including in particular residential mobility and daily mobility, following the pioneering work of Yapa *et al.* (1971). This interpretative approach, attractive because of the interactions it incorporates, is nonetheless complex and difficult to understand for two main reasons: data of this type are rare, due to the methodological difficulties of collecting them in the form of a longitudinal study analysing daily movements as a function of changes in residence, but also because the current paradigms of mobility do not provide adequate conceptual tools to study these interactions. However, certain key concepts allow specification of what is meant here by local mobility. For this, three stages are necessary.

The first is that we use the term "spatial mobility" only for persons moving within a defined geographical space. We recall here the distinction maintained by Kaufmann (2005) between, on the one hand, a movement in space and, on the other, movement within a certain social scale. In the second stage, this principle of spatial mobility treats the latter as being a potentiality and/or a skill. Thus, depending on their social characteristics a person might move, or not. Spatial mobility thus becomes social-spatial mobility, or, more simply, geographical mobility. Finally, in the last conceptual stage Kaufmann's (2000) typology provides a key distinction between different geographical mobilities according to spatio-temporal criteria (Table 5.1).

Table 5.1 – The different geographical mobilities

	Movement within a services and working area	Movement towards the outside of a services and working area
Cyclical movement	Daily mobility	Travel
Linear movement	Residential mobility	Migration

Source: Kaufmann, 2000.

The temporal dimension contrasts cyclical movements that involve regular repetition and imply a return to the point of departure, with linear movements that do not necessarily involve a return journey. The first time category relates to daily life and the second to the life cycle, which is the source of many residential movements and migrations (see Lord and Gerber, Chapter 6). The temporal dimension distinguishes between movements internal to a services and working area, defined as the "smallest area within which the inhabitants have access to both facilities and employment" (INSEE 2003). This area is thus a space offering the potential for activities (including employment) and accommodation to individuals, whether or not they frequent these places in their daily lives. It is clear that this definition, which will also be used as the basis for the empirical analysis of our system of local mobility, voluntarily excludes residential migrations[1].

Using this typology based on the spatial and temporal dimensions of movements, daily mobility is treated here as the totality of movements that are activities of daily life (work, leisure, shopping, visits…), the scope of which are restricted (in time and in space) by two criteria:

- the distance between the location of the activity and the home must be less than around 100 km (if fast modes of transport are involved);
- the activity must be repeated regularly and involves behavioural routines.

Residential mobility, on the other hand, is distinguished from residential migration in that the latter is associated with a relocation due to a work opportunity in another region, possible heliotropism (of pensioners, for example), or due to difficult living conditions in the country of origin (political refugees, for example). Residential mobility in contrast relates more to the life cycle (birth of a child, wedding, etc.), the local

[1] We do, however, provide some discussion of these.

housing markets, or to reasons relating to a neighbourhood or surroundings, and is thus defined by residential strategies (Bonvalet and Fribourg 1990). Certain authors also mention the importance of spatial capital, which is acquired over the course of time in a place and which cannot be physically transferred to another location that is some distance away (Lévy 2003, on Luxembourg see Carpentier and Gerber 2009). This strategic limitation implies that residential mobility always functions within a certain social, economic and cultural continuity.

Thus defined, local mobility is a powerful factor in the organisation of space, a major social influence that often impacts on environmental issues (Newman and Kenworthy 1999) or on issues relating to residential communities (Maurin 2004). However, daily mobility and residential mobility are still more commonly studied separately than as interrelated issues. Peri-urban zones, and in particular those of Luxembourg and its surrounding areas, are ideal for the analysis of this type of interaction. Within a cross-border context, we focus on both the movements between home and work, which are essential components of the services and working area due to their frequency and low flexibility, and on the residential movements of those in active employment.

Towards cross-border peri-urbanisation?

The study of a services and working area is difficult or even impossible according to the National Institute of Statistics and Economic Studies criteria (Institut national de la statistique et des études économiques, INSEE), (INSEE 2003) in the specific geographical context of the Greater Region, where it is difficult to compare and standardise the necessary data for the definition of the region in question. It is, however, possible to overcome this problem by focusing on those in active employment in Luxembourg, making use of methods including the use of the data from the General Inspectorate for Social Security (Inspection générale de la sécurité sociale, IGSS), as well as data collected in studies over recent years.

Differentiated mobility practices

Despite the biases introduced by counting jobs at the administrative headquarters of organisations, which leads to the number of jobs in Luxembourg City being overestimated by around 16% (Sohn and Walther 2008) at the expense of other municipalities, the IGSS data nonetheless allows the temporal dimension of the system of mobility to be measured, including both cross-border commuters and residents. Between 2005 and 2008, of the some 260,000 people in active employment on these two dates almost a fifth changed their municipality of residence (Figure 5.1).

Figure 5.1 – Residential mobility of those in active employment in 2005 and 2008

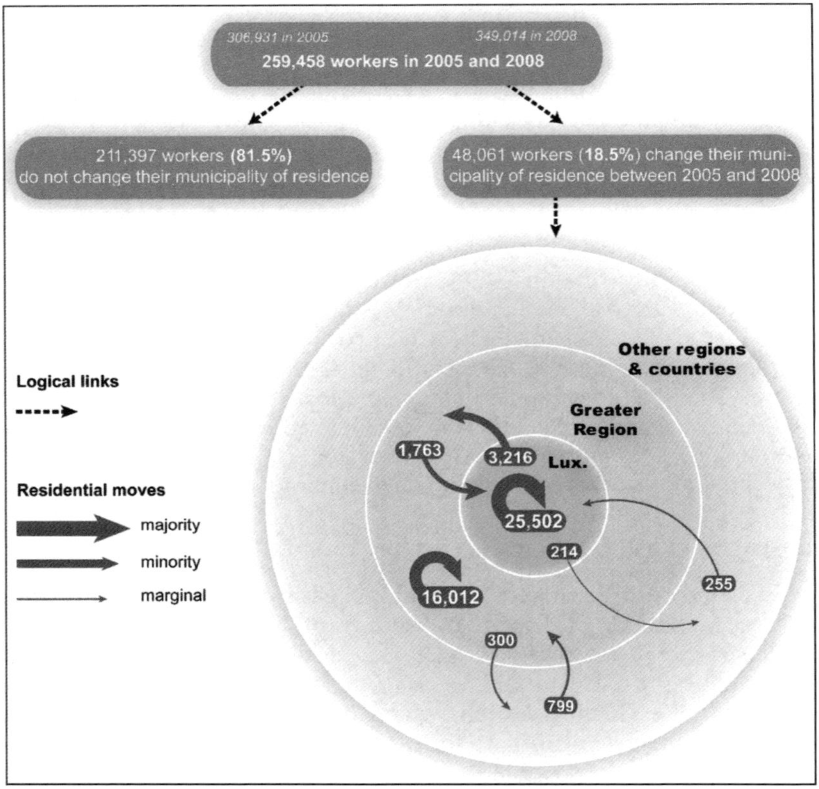

Sources: IGSS 2005 and 2008.

Authors: Gerber, Carpentier and Klein, 2009.

This mobility, which can take place independently of borders, nonetheless shows clear preferences in favour of the region of origin. For example, of the 11,000 in active employment from Lorraine who changed their place of residence between 2005 and 2008, 90% remained in Lorraine. The percentages are roughly the same for the other border regions within the Greater Region (Rhineland and Saarland: 83%; Walloon Region: 79%), while overall almost 90% of those in active employment in the Greater Region remained in their region of residence over this period. This situation indicates a relatively high level of territorial attachment for the border population, whether in favour of the country or the region of origin (Enaux and Gerber 2008). The same applies to workers living in Luxembourg, as the vast majority (88%) remain resident in their own country.

These findings conform with what can be observed for the residential migration of those in active employment living outside the Greater Region in 2005 and returning to Lorraine, Saarland/Rhineland or the Walloon Region in 2008. For example, in the case of French residents, 82% of those in active employment working in Luxembourg and living in France outside of Lorraine in 2005 would be living in this region in 2008. These figures take into account only certain residential migrations, primarily those associated with the attractiveness of Luxembourg and in particular its capital. As a result, these migrations, minor in the context of our cohort of those in active employment in 2005 and in 2008, will be of greater significance when looking at the new arrivals on the Luxembourg employment market. Many of those in active employment thus prefer to be located directly in Luxembourg, especially highly qualified persons and those originating in countries outside the neighbouring countries. As an illustration, of those Europeans[2] in active employment who found work within the last six months of 2008, over 50% moved to Luxembourg. In contrast, of those Belgians in active employment who worked in the same period, less than 10% moved to Luxembourg.

These figures are confirmed by the study carried out by CEPS/INSTEAD in 2003 on Luxembourg's cross-border workers. Over 80% of the latter spent the greater part of their lives less than 80 kilometres from the borders of Luxembourg before coming to work in that small country. Of these "native" cross-border workers, according to Berger (2005), three-quarters have completed secondary education. A majority of the "adoptive" cross-border workers, who are much less numerous (less than 20%), are characterised by having a place of origin over 80 km from the borders (although they have moved closer to the borders over time), have a post-secondary education, which corresponds more closely to the profile of urban employment. These figures put into context the size of residential migration, which is certainly significant within Luxembourg for certain key nationalities[3], but remains marginal within Luxembourg's borderlands.

However, the metropolitan workers (Sohn and Walther 2008) who cross a border to live in Luxembourg are not necessarily closer to their place of work than others in active employment who live in Luxembourg,

---

2 For statistical and legal reasons, it is EU citizens in active employment excluding the French, Belgians, Germans, Luxembourgers and Portuguese who are considered here.

3 This concerns primarily EU citizens, with other nationalities representing only around 10% of the foreigners employed in Luxembourg.

despite a heavy concentration of the places of work of the former (see Figure 5.2). Whatever the distance or time travelled, there is no difference between blue-collar workers and white-collar workers, being around 19 minutes and 14 kilometres for both groups for a single journey. These characteristics of daily mobility vary to a greater extent for cross-border workers. Company managers live further from their place of work than blue-collar workers, with medians of 45 and 40 km respectively between home and workplace, representing journey times of 45 and 40 minutes.

Figure 5.2 – People in active employment in 2005 and in 2008 and place of work

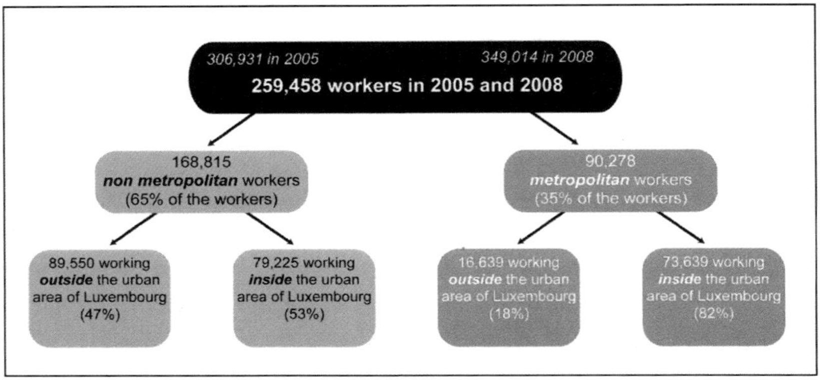

Sources: IGSS 2005 and 2008.

Authors: Gerber, Carpentier and Klein, 2009.

This situation implies that many cross-border workers travel significant distances, with half of them having to travel over 40 km to reach their place of work, usually by car. This involves daily tensions that will lead at least some cross-border workers to seek a home closer to their workplace. One can suppose that these workers experience residential mobility in which the borders play a dual role:

- that of "barrier", due to differences in real estate prices
- that of "attractor" because a move across a border would allow the journey to work to be reduced.

The increase in those in active employment in the metropolitan area over recent years alone promotes the progressive removal of this barrier, by increasing cross-border residential mobility in order to be in closer proximity to places of work, which are mostly located in the capital and its agglomeration. This hypothesis, which has been partially confirmed in statistical terms (Pigeron-Piroth 2008), will now be analysed using the totality of local mobility flows of those in active employment.

A marked peri-urbanisation

First of all, the residential dynamics of the Luxembourg cross-border employment area can be examined by using the relative balance of arrivals and departures for each district. The resultant net migration (Map 5.1) shows a clear structuring of the space, with attractive (red) zones and zones with a negative balance (in blue).

Map 5.1 – Net migration within the cross-border metropolitan region of Luxembourg, 2005-2008 (see colour section)

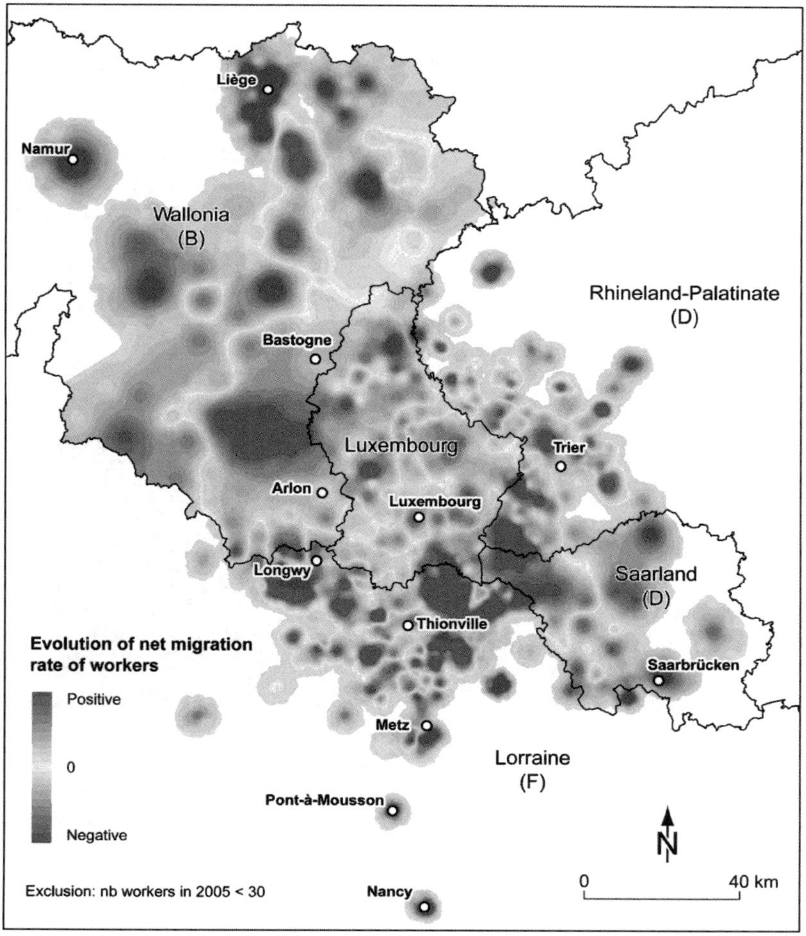

Sources: IGSS 2005 and 2008.

Authors: Gerber, Carpentier and Klein, 2009.

When examining the map, the first point to note is that there is a dichotomy between the dense urban zones and the peri-urban zones. The former have negative net migration rates, while the second in contrast are attractive, for all the countries examined. This first finding clearly illustrates the process of peri-urbanisation in Luxembourg's employment area.

In our area of study, this (peri) urban dynamic is clearly focused on the borders. The map illustrates the strong attraction of less dense zones on either side of Luxembourg's borders. This is the case in particular in the Moselle valley, around Schengen, to the west of the Oesling between Arlon, Bastogne and Mertzig, and, to a lesser extent, in Pays-Haut Lorrain. The areas further away from the border, however, tend to see net departures, acknowledging the importance of the time costs of movement in residential choices[4].

Although the dense urban zones have, with the exception of Trier and to a less extent Arlon, negative local residential flows, this negative balance in fact masks considerable differences in terms of arrivals and departures (Figure 5.3).

Figure 5.3 – Residential balance of main towns, 2005-2008

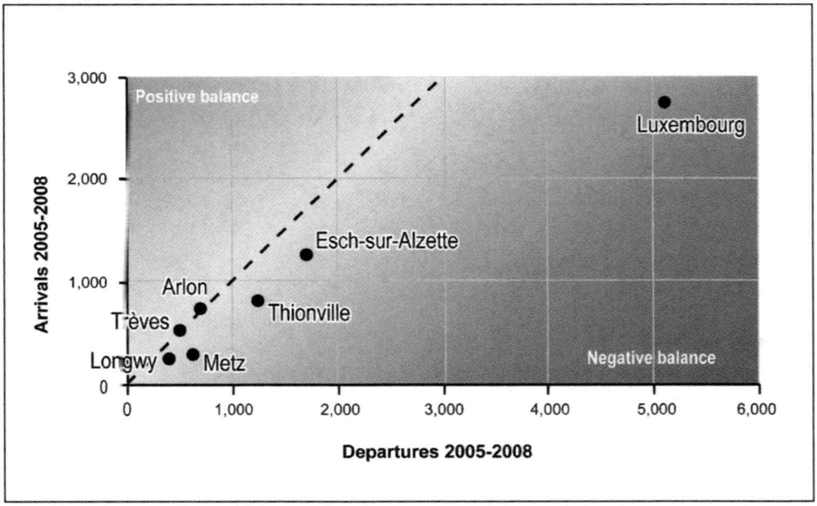

Sources: IGSS 2005 and 2008.

Authors: Gerber, Carpentier and Klein, 2009.

4 These analyses are confirmed by those carried out by INSEE for Lorraine (CES Lorraine 2009).

Luxembourg City has in this respect a unique profile in terms of the size of residential flows of those in active employment, which is larger than for the other cities. The Luxembourg capital is thus seen to be a place with a very high level of residential mobility. This atypical profile shows both the repelling factors, namely the housing prices, which are not comparable with that in other cities in the area studied, and attraction factors, in particular the good availability of employment and the urban amenities specific to Luxembourg City. Undoubtedly, those arriving and those leaving have different profiles. Linking departures to the property market, it can be assumed that the life cycle and, in particular, the home ownership desire, "pushes" certain young households towards the peripheral or border areas (Carpentier 2010). In contrast, the individuals arriving in the capital are either the wealthy classes, in particular those belonging to the international community (see Lord and Gerber, Chapter 6), whose wealth is commensurate with the real estate prices, or young people in active employment who rent accommodations.

The other cities in the Luxembourg functional region have relatively similar profiles, with slightly negative or, for Trier and Arlon, neutral balances. The differences can be explained by both the heterogeneity of the administrative divisions and different dynamics[5].

While the initial analysis of the residential balances has provided evidence of cross-border peri-urbanisation, we will now analyse the flows that support this process. The orientation and intensity of these flows will allow a better description of the spatial mobility interactions that give rise to these net migration rates.

Border effects and local mobility

Given the complexity of the interactions between daily and residential mobility in the Luxembourg cross-border metropolitan region, and following the approaches of Tobler (1976) and Taylor (1983), the approach taken here will be based on a cartographical analysis.

A cartographical approach to border effects

In seeking to understand movements spatially, independently of their nature, it is essential, on the one hand, to make use of both the totality of the locations of origins and destinations, and, on the other, a matrix of exchanges, generally square, containing the attributes of each pair of

5 As such, the net migration of Arlon, which is to a slight degree a beneficiary, essentially reflects the peri-urban dynamic within a single district containing multiple peripheral areas of an undeniably rural nature. To a lesser extent, this applies also to Trier.

movements between the places in question. At first sight, the problem to be solved appears relatively simple: "*Knowing the locations of n places and the movements made between them, how can these movements be represented spatially?*"

The most common cartographical method is to present the movements between each pair of places in the form of a flow. According to the rules of graphic semiology (Bertin 1967), use of the visual variable of "size" is the most commonly used solution, which shows visually the orientated links, with a thickness proportional to the quantity of movement. This relatively old approach, derived from the first attempts made by Harness in 1837 and Minard in 1845 (cited in Palsky 1996), remains today the most commonly used cartographic solution. Although essential, these maps are relatively limited in most cases. When there are too many places, the results become illegible, due primarily to the high visual density, with the number of potential movements increasing with the square of the number of places between which relations exist (Klein 2007). Thus, for n places there are n (n-1) orientated links, which means that for the 116 municipalities of Luxembourg there are 13,340 potential links to be represented.

Given this significant mass of information to map, it is essential to use processes that allow the amount of information to be represented to be reduced. Taking a descriptive approach, the most common procedure consists of reducing the number of flows either spatially (by reducing the size of the zone being studied or aggregating places) or thematically (by suppressing movements or orientations, or by introducing minimum levels). These options, although effective, do not allow the systematic resolution of all problems of legibility, such as those of superimposition, which may persist. In addition, the selection of flows is usually done at the expense of the smaller ones. The flows linked to the phenomenon of peri-urbanisation, which are often spread across a large number of spatial units, may be eliminated from the cartographic representation. To avoid this difficulty, Tobler's modelling approach synthesises all the flows, including the smallest, using vector calculations (Tobler 1976 and 1978). Three categories of map can then be constructed (Tobler 1982, Cauvin and Reymond 1991). *The first category* (Tobler 1981) corresponds to a type of cartography where one shows for each place the result of movements from this place to all others, weighted in terms of all exit movements. This result being a vector, the maps obtained can be considered as irregular vector fields, with the points of origin of the vectors being distributed irregularly. *The second type* is a representation of the same data as before, but in the form of a regular vector field. In this case, the data is distributed over the whole of the zone studied using interpolation. Finally, in the *third category*, the data obtained previously can also be represented by continuous lines that favour the representation of general directions over local features.

Choice of method and application to the Luxembourg cross-border metropolitan region

Applied to the local mobility system of the cross-border metropolitan region, the method proceeds in two successive steps with, firstly, the aggregation of the spatial units and then a mathematical procedure that allows identification both locally and globally of the main directions for each category of movement.

The first stage consists of homogenising the spatial units. With the district units being of different sizes across the Greater Region, it is necessary in this first stage to homogenise these spatial units. On the basis of the MAUP (Modifiable Areal Unit Problem) methodology (ESPON 2006), a more appropriate division, based on the aggregation of the elementary units, can be designed for the whole of the Greater Region. The outcome is produced on the basis of two fundamental criteria of the MAUP method – equal population and functional areas – in order to guarantee better homogeneity of the new spatial units created.

Taking into account this first stage of homogenisation, a consistent representation of the movements is now possible for the zone of study, without interpolation being required. As a result, in the second stage, a map based on the resultant vectors[6] is entirely suited to allowing visual comparison of the different types of movement (here, daily and residential movements) or of movements taking place at different times. This approach allows in particular both the most common directions and the volumes of flow, at their place of origin, to be shown all at once (see Figure 5.4). Under these conditions, as elaborated by Cauvin *et al.* (2007), the map allows spatial characteristics that are invisible or little known to be shown visually.

6 Thus corresponding to the first category of vector representation proposed by Tobler.

Figure 5.4 – Methodological insert

The resulting vector method requires:

(i) the transformation of the flows into unitary vectors whose direction is given by the line from the point of origin to the point of destination, and the norm is equal to the sum of the flows in question.

(ii) the calculation of an average angle using the formula below:

$$\text{AngleM} = \arctan\frac{\sum_{i=1}^{n}(\sin\theta_i)}{\sum_{i=1}^{n}(\cos\theta_i)}$$

AngleM = M-Angle

where θ_i represents the angle formed by vector i

The final value of this angle of the resultant vector requires an additional adjustment as a function of the quadrant in which it appears, in accordance with the conditions below:

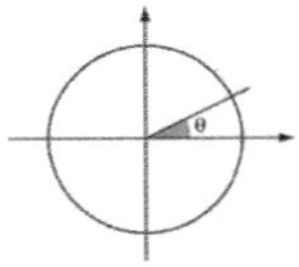

$\sum_{i=1}^{n} \sin\theta_i > 0$ *and* $\sum_{i=1}^{n} \cos\theta_i > 0$*, then MeanAngle = AngleM*

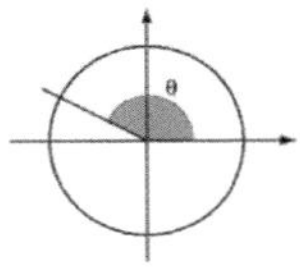

$\sum_{i=1}^{n} \sin\theta_i > 0$ *and* $\sum_{i=1}^{n} \cos\theta_i < 0$*, then MeanAngle = 180° - AngleM*

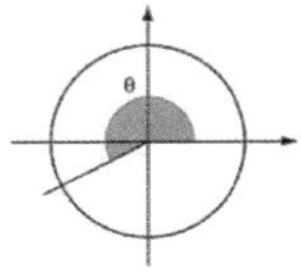

$\sum_{i=1}^{n} \sin\theta_i < 0$ *and* $\sum_{i=1}^{n} \cos\theta_i < 0$*, then MeanAngle = 180° - AngleM*

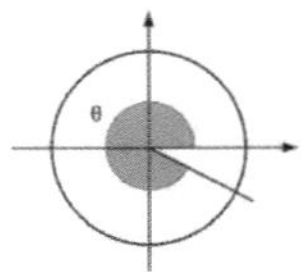

$\sum_{i=1}^{n} \sin\theta_i < 0$ *and* $\sum_{i=1}^{n} \cos\theta_i > 0$*, then MeanAngle = 360° - AngleM*

(iii) the calculation of the quantity of movement, written Fi, calculated from point of origin, is equal to the sum of the totality of movements exiting the spatial unit in question:

$$F_i = \sum_{j=1;j\neq i}^{n} F_{ij}$$

Now, the thematic interpretation of the maps produced allows a better understanding of the local system of mobility in terms of movements between home/work and the residential mobilities.

Highly focused daily mobility

Map 5.2 – Main orientations of home/work flows, 2008

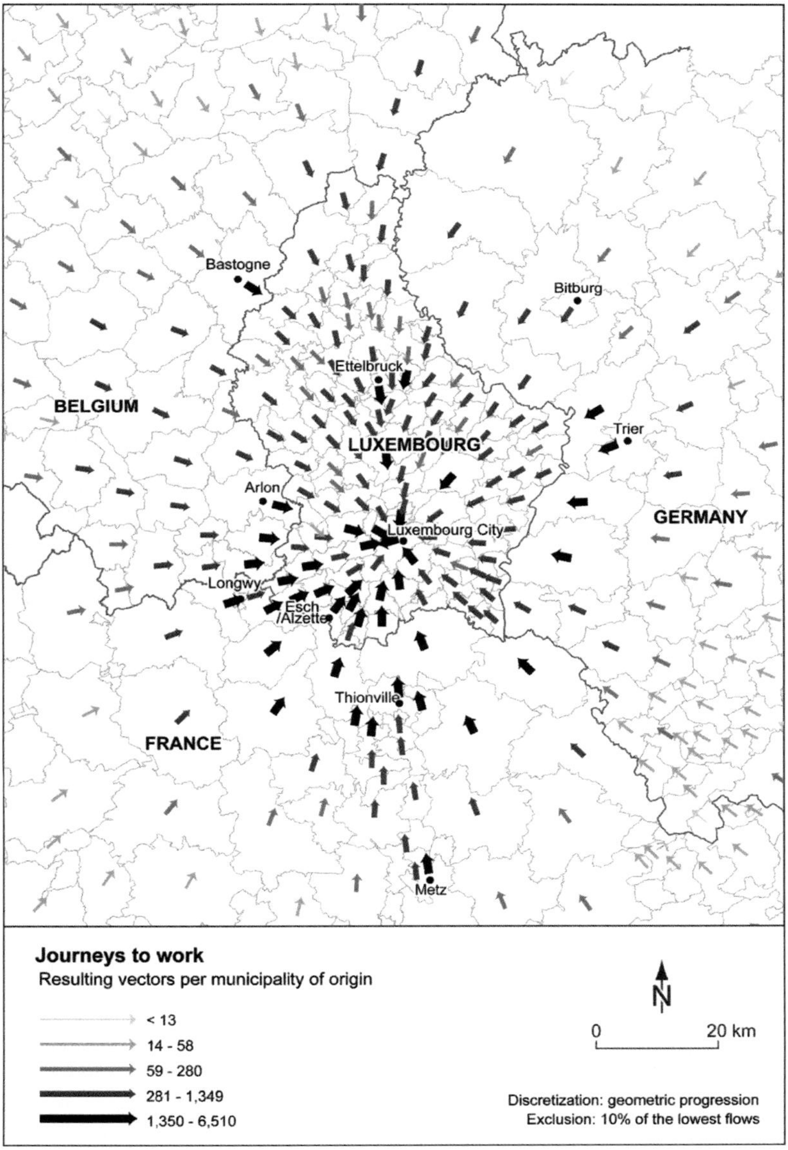

Sources: IGSS 2008.

Authors: Gerber, Carpentier and Klein, 2009.

The analysis of the home/work mobility of those in active employment in Luxembourg (Map 5.2) shows two major elements. First, there is a gradient of intensity from the centre to the cross-border peripheries, a result that was to be expected as it was taken into account in the delineation of the area of study. Only the Moselle valley deviates from this gradient, due in part to its easy accessibility via motorway and train, and in part to its high population density. The map also shows a strong convergence of average flows towards Luxembourg City, originating both in Luxembourg and in the Greater Region, which demonstrates the major role of the Luxembourg capital in structuring the employment area and thus the services and working area.

Given this strong concentration of employment, it is now necessary to analyse the residential dynamics of those people.

The border – a residential attractor?

Examining the local residential movements in terms of average flows (Map 5.3) is a useful complement to the first analysis, carried out in terms of balance (Map 5.1).

An initial examination shows that there are two complementary dynamics. First, there is a border effect in the form of a "barrier effect" applying to residential flows. It is seen that for all the countries considered, the flows originating in border districts are not orientated towards the other side of the border. Despite the existence of cross-border residential flows (Pigeron-Piroth 2008), the latter play a less significant role, in quantitative terms, in the structuring of the totality of the movements of people in active employment. The strong differentials in terms of real estate prices on either side of the borders and the effect of nationality, particularly in terms of place attachment demonstrated by Enaux and Gerber (2008), do not therefore appear to be contradicted by these aggregated analyses.

Map 5.3 – Main orientations of residential flows, 2005-2008

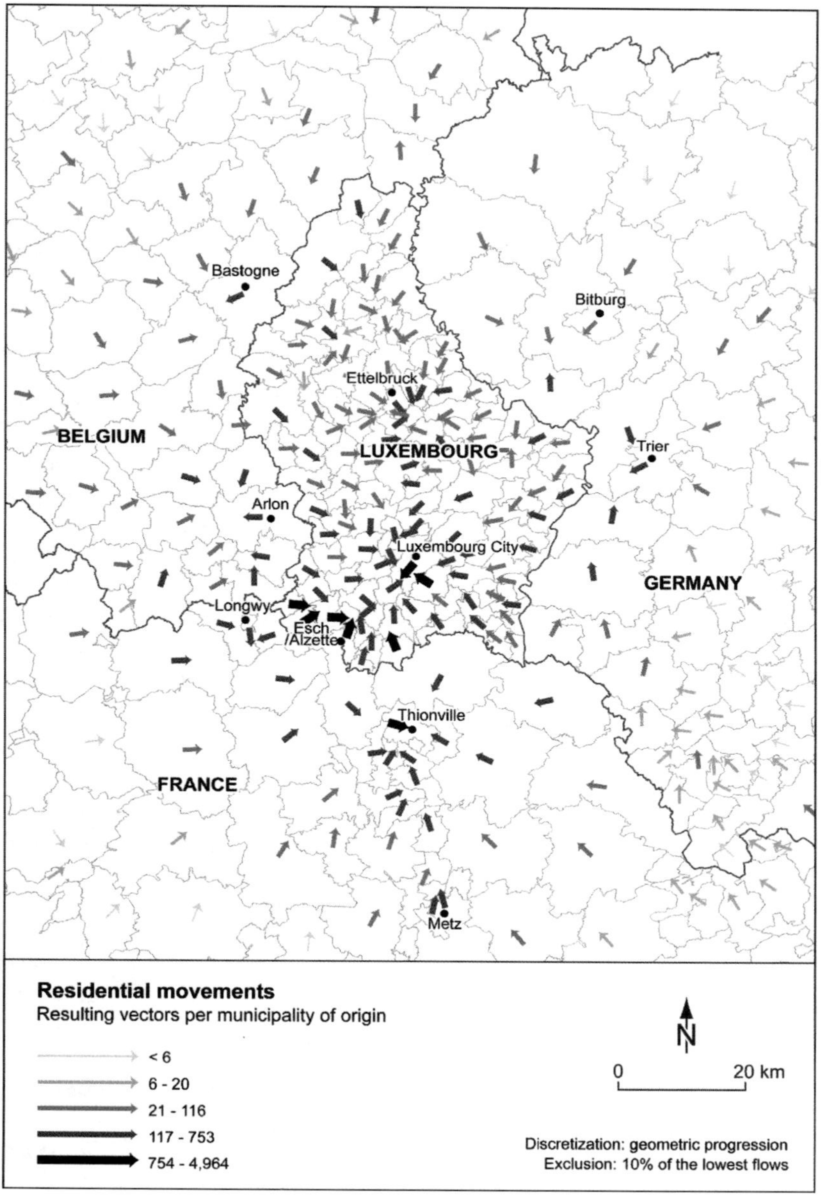

Sources: IGSS 2005, 2008.

Authors: Gerber, Carpentier and Klein, 2009.

The second element structuring the residential dynamics is the strong attraction of peripheral districts close to urban centres. The majority of average vectors, representing the direction of residential flows, are towards urban zones; it should be remembered here that the map of net migration shows that the dense urban areas have negative balances. This tendency towards harmonisation illustrates both the search for improved accessibility and the attraction of urban amenities.

Finally, the cross-border workers residing in the districts that are especially far away from Luxembourg display a uniform tendency to move closer; this illustrates the fact that there is a limit in terms of the time available for daily mobility. In other words, an individual living over 100 km from Luxembourg will have a strong propensity to move closer to its place of work. The extension of the Luxembourg employment area (Schneider and Pigeron-Piroth 2009) has thus perhaps reached its limits in terms of spatial size and its future development will probably be via demographic growth in the districts already involved.

Ultimately, the joint analysis of net migration and average flows demonstrates a peri-urban dynamic that is not cross-border but on either side of the borders. This phenomenon is supported by both the centripetal flows, from the distant border peripheries, and by the centrifugal flows, illustrating a dynamic of urban sprawl. The totality of these processes defines an employment area marked by strong local mobility but whose spatial extension appears to have reached its limits, as transport costs, both in terms of financial cost and the time required for movement, are increasingly important in residential decision making.

Conclusion

The ongoing process of metropolisation in Luxembourg and in its border regions is accompanied by a strong process of peri-urbanisation on either side of the country's borders. The concentration of high added-value activities in the capital is leading to several major spatial processes. First, at the local level, the employment area has expanded greatly, even if those employees furthest away are now tending to move closer. Next, although the capital has a negative balance in terms of local residential mobility, it is nonetheless the main point of entry for immigrants (of all socio-economic profiles), which explains the fact that, overall, Luxembourg City is experiencing a strong positive demographic growth.

The articulation of these two levels, local and global, is at the heart of the process of metropolisation. It results in both a selective distribution of those in active employment in accordance with their socio-economic profile, their country of origin and probably their date of arrival in the country, and a certain levelling of residential behaviour, so that the trend towards peri-urbanisation appears to be dominant in the four countries of our zone of study. Looking beyond the negative externalities of peri-urbanisation on the environment, already much discussed in the literature (Newman and Kenworthy 1999), this article examines the social dimension of the flows of local mobility and the territorial cohesion of an emerging cross-border metropolitan region.

The interactions of daily and residential mobility are heavily influenced by the presence of national borders. With regard to home/work mobility, the border plays the role of an attractor. The strong differentials in terms of salaries and the availability of work encourage cross-border workers to come to work in Luxembourg. One illustration of this attraction is the very high level of journey distances and times in relation to the size of the agglomeration of Luxembourg (comparable to those seen in the Paris region). In contrast, at the level of residential movements, the border plays the role both of attractor, as the dominant trend is one of moving closer (at least for cross-border workers), and the role of barrier, illustrated by the relative weakness of flows of households moving across borders. While the differentials can again be cited, the cultural differences cannot be ignored, in particular for those people originating from neighbouring countries.

Finally, all these results raise two main questions for the future development of the cross-border metropolitan region of Luxembourg. First, from a social perspective, it is pointed out that metropolisation could be a risk for territorial cohesion. Indeed, the high concentration of value in Luxembourg City and its urban area generates very high land rents that encourages the less qualified people to move further away from their work places. Furthermore, cross-border differentials can also encourage people to move outside Luxembourg, thus favouring social segregation (Carpentier 2010, Lord and Gerber, Chapter 6). Then, the sustainability of this cross-border mobility system is also in question. Considering the longer distances generated by this cross-border spatial mismatch, cross-border workers are mostly facing strong car dependence. This issue may be tackled by improving cross-border cooperation in the public transportation field. A first initiative is now given by the Transborder Mobility Strategic Scheme (Schéma stratégique de mobilité transfrontalière, SMOT) signed between the Luxembourgish and French authorities in 2009. In this agreement, concerning a common strategy in public transport planning, new train lines and stations, as well as new bus lines, between the two countries are foreseen, in order to favour a modal switch towards public transport. These projects reflect the desire to strengthen the sustainability of daily mobility within the cross-border metropolitan region.

Map 1.2 – Location of jobs linked to the knowledge economy in Luxembourg City, 2005

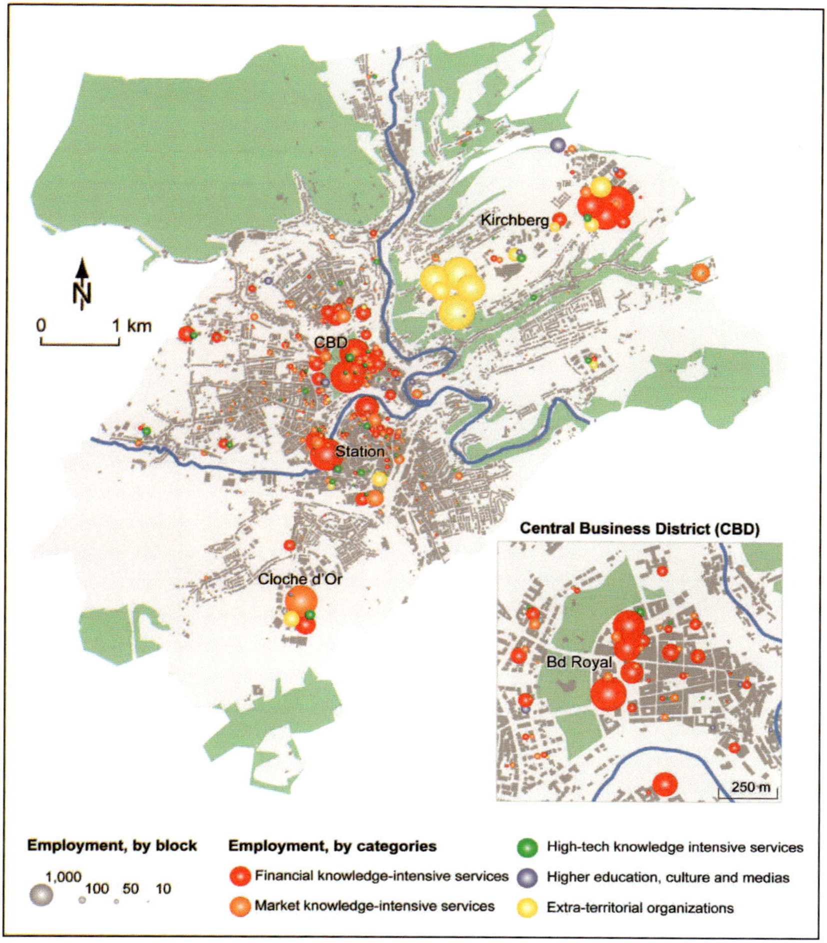

Source: STATEC

Cartography: Sohn (2009b), modified by Walther.

Map 2.1 – Location of the financial establishments in Luxembourg, 2005

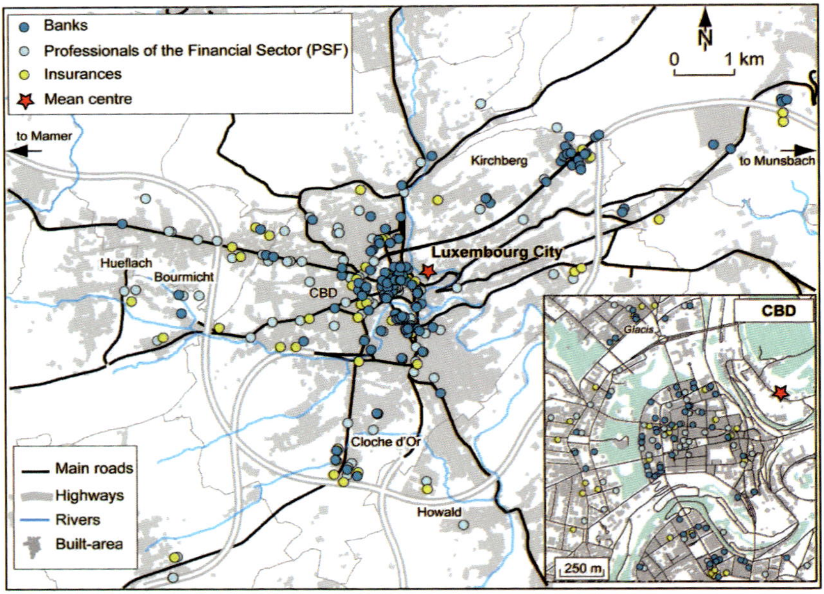

Source: CSSF. Cartography: Walther.

Map 4.2 – Density of commuters working in the agglomeration of Luxembourg by place of residence, 2008

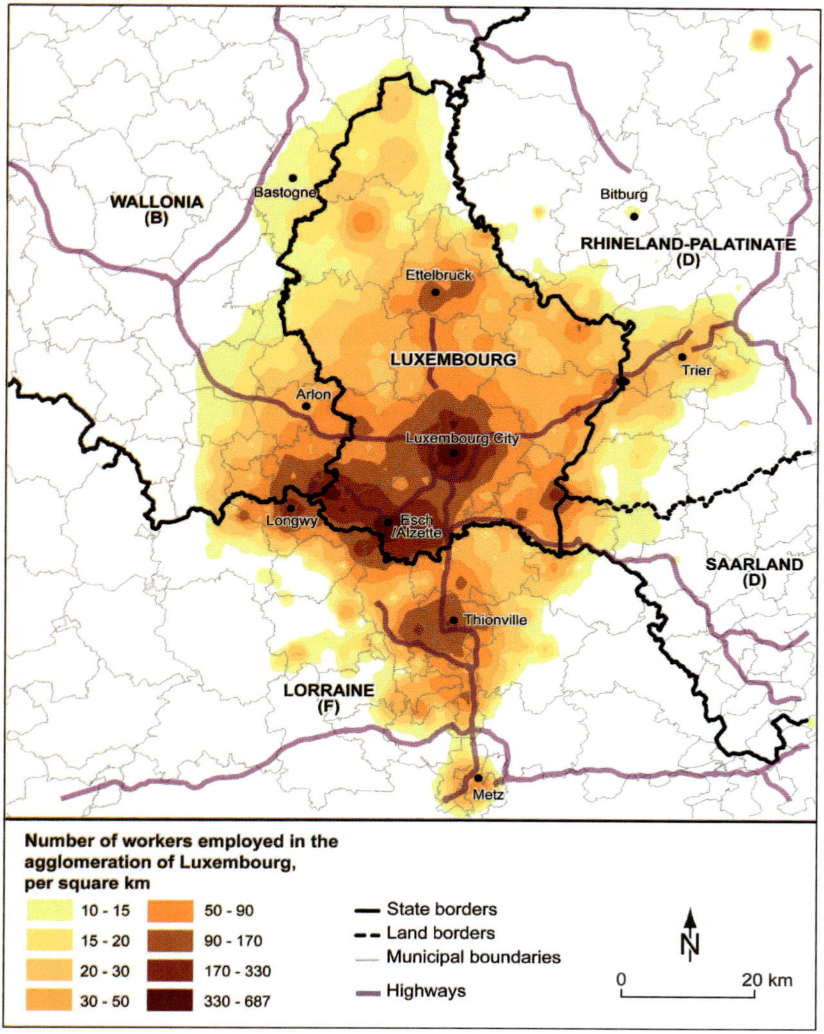

Sources: IGSS 2008.

Authors: Decoville and Sohn, 2009.

Map 4.3 – Spatial expansion of the area of influence of the agglomeration of Luxembourg, 1994-2008

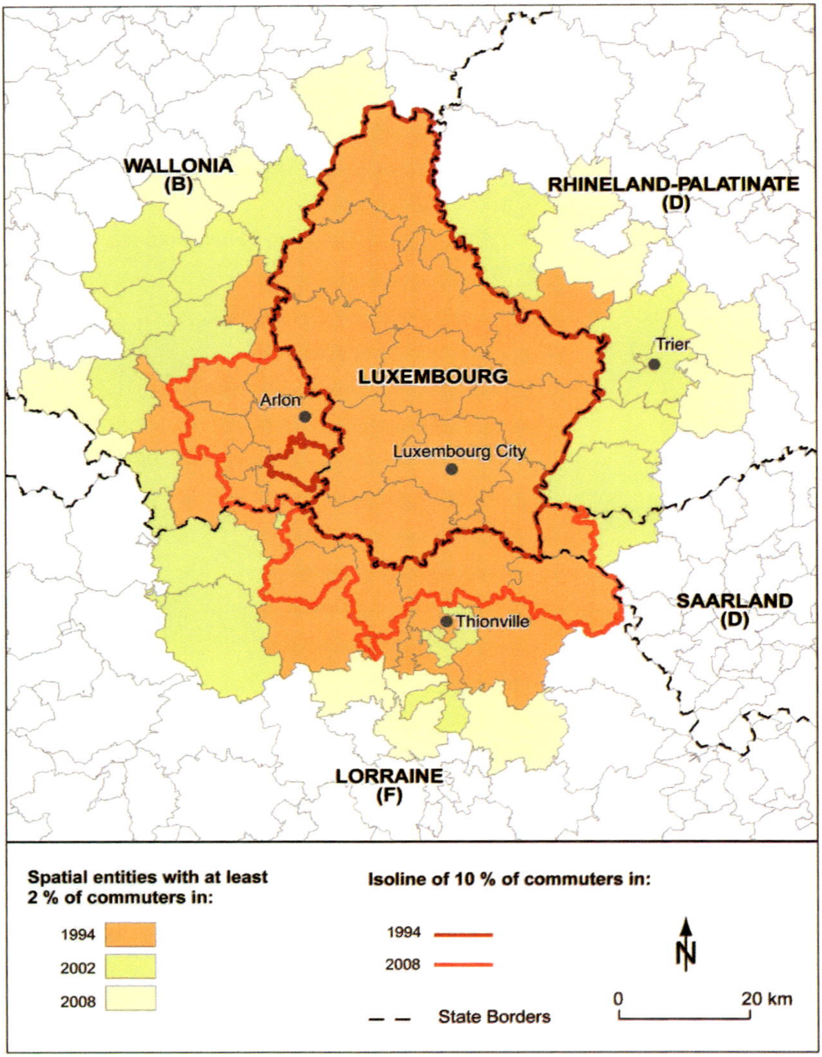

Sources: IGSS 1994, 2002, 2008.

Authors: Decoville and Sohn, 2009.

Map 4.4 – Relative changes to density of commuters to the Luxembourg agglomeration, 1994-2008

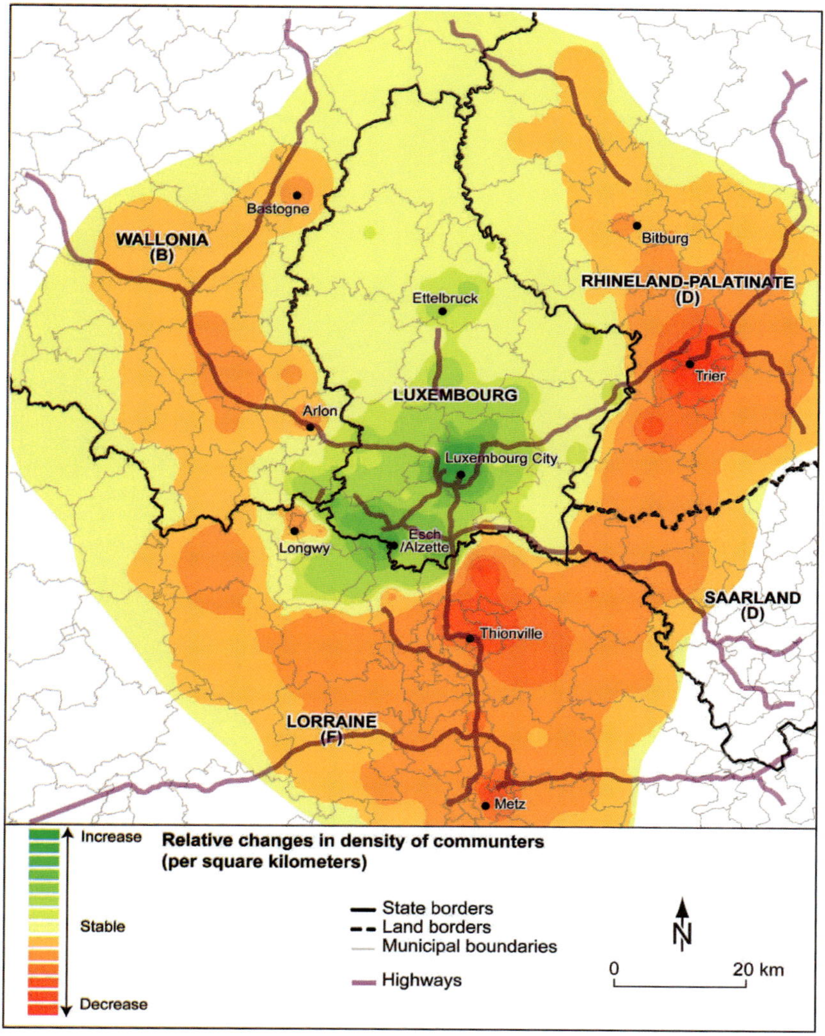

Sources: IGSS 1994, 2008.

Authors: Decoville and Sohn, 2009.

Map 4.5 – Place of residence of metropolitan commuters, 2008

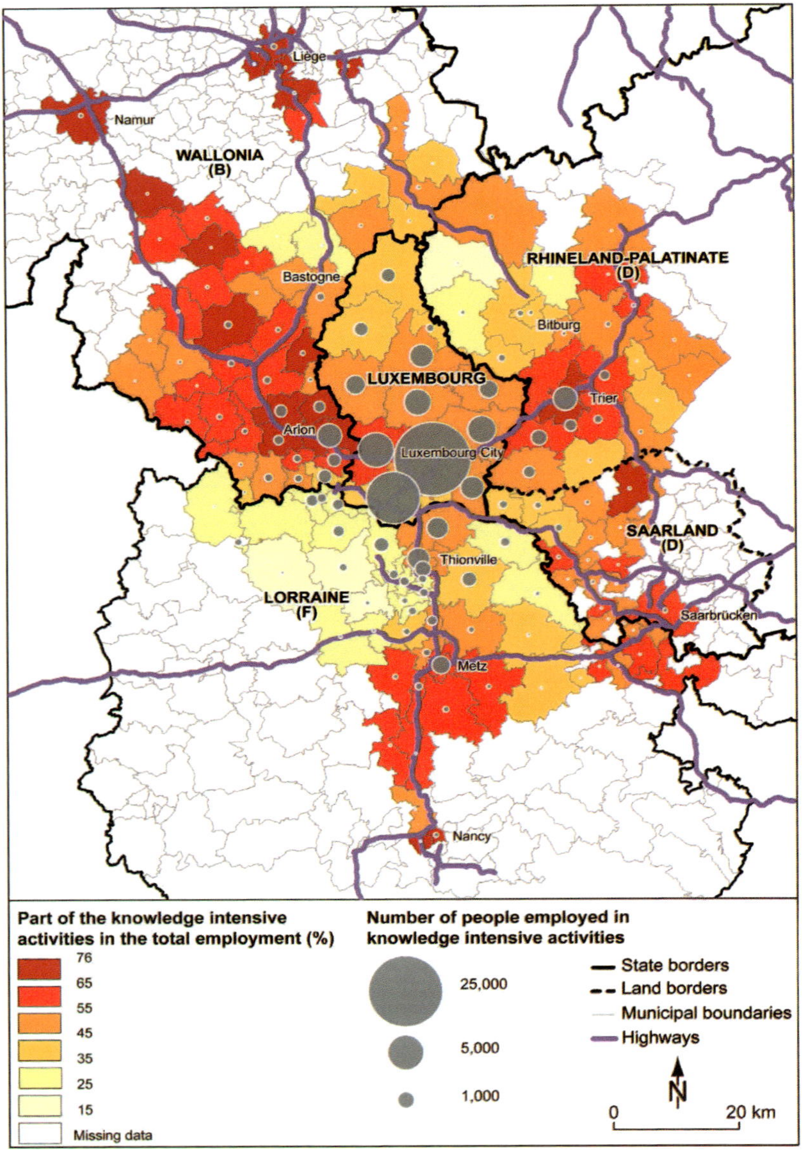

Sources: IGSS 2008.

Authors: Decoville and Sohn, 2009.

Map 9.1 – The Greater Region in its European context, 2003

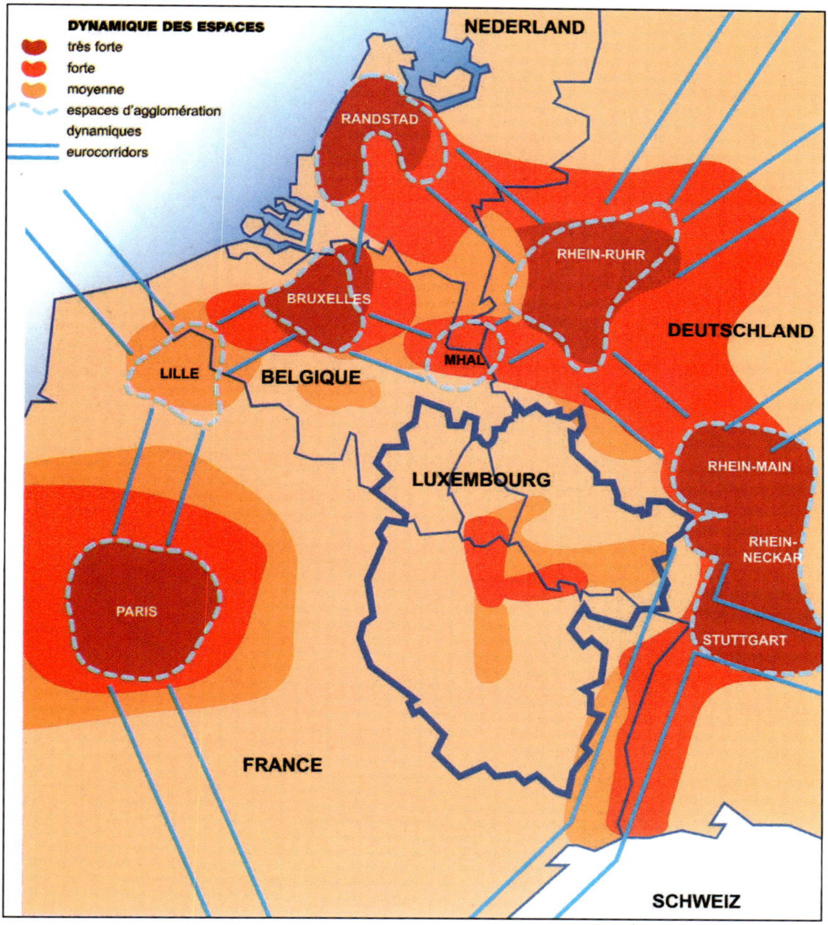

Sources: Ministry of Spatial Planning, Programme directeur d'Aménagement du territoire, Luxembourg, 2003.

CHAPTER 6

Residential and Cross-border Mobility

A Catalyst of Social Polarisation?

Sébastien LORD and Philippe GERBER

Metropolisation is now a reality in Luxembourg – a phenomenon of sprawl, functional diversification and territorial specialisation that can be observed across the borders of the country. Notable features are that the Luxembourgish metropolis is developing around a small agglomeration and its capital, Luxembourg City, is located close to the borders with neighbouring countries. The area of attraction of this metropolitan region thus extends well beyond the national borders (Sohn and Walter 2009), a fact made visible by the 130,000 cross-border commuters (Carpentier and Gerber 2007).

Luxembourg has been a country of immigration for several decades, with over 40% of its residents now foreigners (Burnotte-Boreux and Casteels 2003). For a long time, immigrants to Luxembourg were low-qualified working class people, in particular Italians and Portuguese (STATEC 2009a). Their arrival made up for a lack of workers in certain key sectors, such as the mining and steel industries. Today, still within a framework of providing labour in strategic sectors, the development of service, financial and research activities, as well as those linked to the European institutions, is resulting in the presence of highly qualified immigrants, a "gilded" group often referred to as a transnational class (Sassen 2001, Weiss 2005).

The acceleration of metropolisation seen over the last decade, and the geographical position of the Luxembourgish capital, have given this process a specific character, in particular the exceptional demographic growth. Luxembourg is a useful case to analyse in relation to possible socio-spatial inequalities. The country can be classed as a place of residence "for the privileged" due to the property and land prices, which are higher than in neighbouring countries. Given the double immigration, several questions can be addressed regarding residential location and the residential mobility within the country and its Greater Region. While several studies have examined the socio-spatial structuration of

Luxembourg and its capital (Carpentier 2007, Lord and Gerber 2009), there are few detailed findings available at the level of the cross-border area of influence.

Metropolisation: from specialization to segregation

Metropolisation is a complex phenomenon, related to concepts of sprawl, fragmentation, connectivity and mobility (Bassand *et al.* 2001). It can be studied at the macro level, with reference to forces of market globalisation, links and exchanges within networks of global capital and the influence of economic, financial and political actors (Sassen 2001). It can be observed that the areas of influence of cities are expressed through strong effects on employment, typically in the tertiary sector, and in one or more strong geographical centres that are *structured in networks* (Dupuy 1991). The residential, industrial and commercial functions competing at the periphery of the centres vary unevenly in accordance with the gradients of decline of greater or lesser intensity.

The metropolitan territory is difficult to understand in terms of morphological, functional and, in the case that interests us, residential structures, especially when it is an international region. Often associated with peri-urbanisation or suburbanisation, metropolisation is always synonymous with high levels of mobility and sprawl (residential and/or urban), (Roux and Vanier 2008), with territorial management becoming increasingly complex. Metropolitan residential areas struggle to implement effective institutional and democratic structures when their development requires joint control of material spaces (Ascher 2001) as well as immaterial (Dupuis 1991, Urry 2003) and symbolic (Castells 2003, Giddens 1981) realities. However, metropolisation in Luxembourg is characterised more by a high level of functional integration of peripheries (national and borders) and less by typical sprawled morphologies.

At the social and cultural levels, the metropolitan territory tends to become specialised and segregated (Da Cunha and Both 2004, Benhamou 2004), creating zones with a high concentration of high-value activities offering betterment to privileged residential spaces close to infrastructure, facilities and services. During this phase, other territories are excluded from all dynamism or possibility of improvement (Sélimanovski 2008, Maurin 2002). As emphasised by Ascher (2003: 612-615), metropolitan urban structures are the privileged sites of new social inequalities. In these contexts, segregation achieves new heights, sometimes with ghettos that one cannot leave and gated communities that one cannot enter (Haumont 2000). These theoretical/empirical frameworks are in fact "extreme situations" that are not found in Luxembourg as well as in border areas. However, these classical approaches allow us to identify and understand segregative trends that could exist.

While an unequal distribution of financial insecurity can be observed across Luxembourg (Gerber *et al.* 2008), the reality of strong social enclaves has not yet reached the country. Lord and Gerber (2009) observe, at the level of the Luxembourgish municipalities, relatively weak ethnic and cultural segregation. That being said, the same studies put forward hypotheses of segregation tending to be based around the upper end of the social scale, structured by both the agglomeration of Luxembourg City and the acquisition of property in relation to the dominant residential model (ownership of a single-family house) across the entire country. In the cross-border context, the key notions of identity (Maurin 2002) and of spatial and social withdrawal, with the development of prestigious residential districts, suggest the hypothesis that such an isolation of poverty takes place beyond the national borders, in the bordering peri-urban areas.

For Luxembourg, one could put forward the idea that the favoured residential zones correspond to the capital and its inner peri-urban ring, as shown by Gerber *et al.* (2008) from a social perspective and Sohn (2006) on a morphological basis. If the Luxembourgish capital can be a privileged residential area at the national level, the whole country could well be its equivalent at the level of the Greater Region. This idea benefits from being extended to the cross-border urban areas located close by and with good transportation links, in particular being accessible to metropolitan cross-border commuters. These areas are, primarily, the southern border with France (Lorraine) and the centres of Arlon in Belgium to the south-east (Wallonia) and Trier and Perl to the east in Germany (Saarland; Rhineland-Palatinate).

Mobility in the metropolitan area

The process of metropolisation brings together overlapping historical, socio-economic, urban and cultural issues relating to the structuration of the socio-spatial environment at the micro level (Castells 2003). Concrete issues concerning individuals' living environments are at stake in its morphological and functional materialisation. While the process resulting in sprawl and in specialisation of urban spaces and resources is closely linked to the ever-increasing mobility of individuals (Kaufmann 2008), it is also linked to the notion of "metropolitan competences" (Bourdin 2005) or the "field of possibilities" (Maurin 2004). Thus, movement within the metropolitan territory is not necessarily easy for all.

The attraction of highly qualified workers to the metropolitan region, and their eventual settlement there, increases the strength and competitiveness of the cities. This class, with high levels of economic, social and cultural capital, is relatively independent in relation to residential space, or even in relation to borders – global citizens (Urry 2003) or transnational classes (Weiss 2005). These migrants, primarily from developed countries in the so-called "North"[1], do not exhibit the main social markers associated with immigration (poverty, exclusion, insecurity). Lower qualified workers constitute the other face of metropolitan growth, particularly those resulting from lifestyles and consumption. While the research on residential segregation and urban ecology has long documented these issues, providing rigorous and valid methodological and interpretive frameworks, the new forms of international immigration and the intensification of metropolitan activities exacerbate these contrasts or contribute to the emergence of new transnational social spaces (Pries 2001).

In a world of movements, choosing a place of residence can be of fundamental importance. The *hyperconcentration* of resources (urban services, schools, places of consumption, etc.) demands skills that are not equally shared by all actors. One can hypothesise that access to the Luxembourgish capital and its close peri-urban surroundings is available only to individuals at a high level on the social scale. While new arrivals often initially establish themselves in a central (and dense) location marked by a social and functional mix, subsequent residential displacements are often towards the periphery for the higher social classes. This has been observed by, amongst others, Da Cunha in Switzerland (2005) and Peach (2001) and Bolt *et al.* (2008) in North America, and Burgers and van der Lugt (2006) in Northern Europe. The new immigrants select central districts when they arrive, but go on to then favour so-called "national" sectors. Without entering into the abundant literature on integration and immigration, or into the debate on assimilation, we emphasise the classic sequence outlined by Park *et al.* (1925) and investigated many times (Musterd and Ostendorf 1998).

Questioning residential mobilities in Luxembourg

Research on residential mobility and lifecycles has extensively documented individual life trajectories. Residential trajectories are fitted into and organised in sequences of events observable across time. Years of research have allowed pivotal events to be identified (end of higher education, first job, departure from family home, arrival of a child), (Bonvalet and Fribourg 1990). These are successive events that affect life trajectories, residential mobility or immigration. Note that this

[1] European Union, Norway, Canada, United States, Japan, Austrialia and New Zealand.

normative approach varies according to culture, era, life context and socio-economic position[2].

In Luxembourg, almost 45% of displacements seen in the most recent available census (2001) are made by individuals aged between 25 and 39 (Burnotte-Boreux and Casteels 2003). Age also appears to be the best explanatory variable for residential mobility in the country. However, differences in employment situations, nationality and family situations are also seen. The unemployed, those with Luxembourgish nationality, young households and single persons appear to be more inclined to move; these sub-groups are also those who move a greater distance. We note that employment results in long-distance residential displacements, or international migrations (Vignal 2006). In Luxembourg, these displacements tend to be more numerous; the preceding chapter indicates a 25% share of total movements (Gerber, Klein and Carpentier, Chapter 5).

The strong economic growth associated with Luxembourgish metropolisation is developing alongside "usual lifecycles" as factors influencing residential mobility within the territory of the Greater Region. This is illustrated in a study by Zanardelli & Brosius (2009) for Luxembourg: while many of those recruited by Luxembourgish companies in 2006 were aged under 30 (43%), a majority of those recruited were over 30, and more than one in four was over 40. How common, then, are these residential movements that are shaking up the patterns of movement normally associated with lifecycles? Are they exceptions or do they constitute a significant portion of residential mobility within Luxembourg and the Greater Region?

Questions and working hypotheses

The first question addressed is that of the magnitude and the direction of residential mobility in Luxembourg and its Greater Region. Specifically, the following questions are explored:

What is the magnitude of residential mobility with the Greater Region of Luxembourg?

a. How many movements are there?
b. What are the origins and destinations of these movements?
c. What proportion of these movements are accounted for by intra-border and cross-border movements?

---

2 Two forces are changing the classic organisation of lifecycles: the increase in the number of family stages (Bonvalet 2005) and the ageing of the population (Quadagno 1999); the latter has added a new period of active life following retirement (Lévy 2006).

Relative residential stability is expected, in particular due to the high price of Luxembourgish property. That said, of the total number of movements, a higher proportion of movements is expected to be within the countries of the Greater Region in comparison to those that are across borders, and more specifically those entering Luxembourg. Daily mobility allows households to make relatively significant changes, contributing to the stability of their residential anchorage. In parallel to such adjustments, and in the context of high property prices, movements towards the Luxembourgish border, within neighbouring countries, are expected to be relatively high in number, in particular non-metropolitan workers located low down on the social scale.

The second issue addressed concerns the socio-demographic characteristics of the individuals making residential movements. Essentially, this involves investigating the factors influencing residential mobility in Luxembourg and the Greater Region.

Which people are involved in intra and cross-border residential movements and also which are concerned by movements into or out of the country or the Luxembourg City agglomeration?

a. What are the nationalities involved?

b. What are the age groups of the individuals moving?

c. What are the salaries and types of employment?

The metropolitan growth in Luxembourg and the Greater Region is tending to change the residential movements associated with lifecycles, in particular with movements not based on the key events usually connected with residential movements. Essentially, it is a case of measuring "inconsistent" movements in order to find out whether residential mobility involves, for example, other categories of individuals than young workers or persons of Greater Region nationalities (Luxembourg, Germany, France, Belgium). Another aspect underlying this second question concerns the socio-economic characteristics of residential movements into and out of Luxembourg and the main urban agglomeration. Movements into the country and its capital may tend to concern primarily foreign metropolitan workers at the top of the social scale. Movements are towards the Luxembourgish border without crossing it, and movements out of Luxembourg and its capital may tend to be influenced by metropolisation but involve individuals at the bottom end of the social scale. The hypothesis upheld here is that there is socio-spatial segregation based around the top of the social scale, which is structuring the Luxembourgish metropolitan territory, with households in less advantageous social positions being excluded from access to the centre, but not necessarily pushed to the outside. Such an adjustment is facilitated by easy access to daily mobility.

Our two hypotheses thus concern four types of residential movement, illustrated schematically in Map 6.1. The first corresponds to the black arrows and concerns residential displacements from the Greater Region into Luxembourg or the agglomeration of its capital. The second type of displacement, shown with white arrows, corresponds to the opposite movements, i.e. out of Luxembourg or the agglomeration of its capital into the bordering countries. The third type, shown with dotted arrows, involves a movement approaching the Luxembourg border but without crossing it. Finally, the last type of displacement, illustrated with round arrows, corresponds to other residential displacements with a bordering country without approaching the border.

Map 6.1 – Four working hypotheses on the residential mobility within the Greater Region

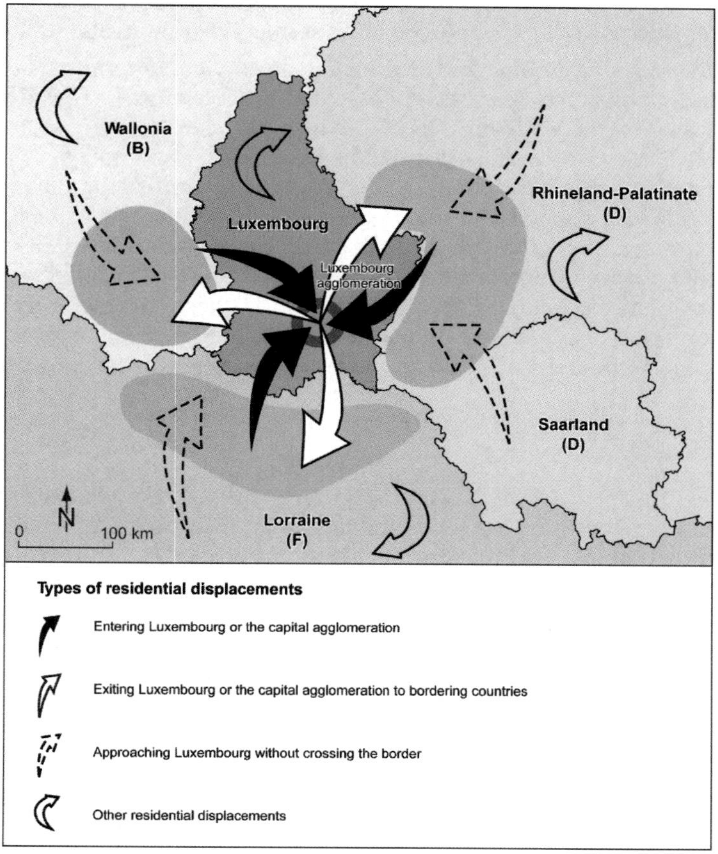

Authors: Lord and Gerber, 2009.

Research and methodological strategy

Many studies have addressed the issues considered here. Few of these, however, look at the whole of a country and a cross-border metropolitan area. The availability of systematic and longitudinal data on the Greater Region, including on individual salaries, thus grants an original dimension to this chapter. The Luxembourgish government has, via the General Inspectorate for Social Security (Inspection générale de la sécurité sociale, IGSS), an individual and longitudinal administrative record of the population resident in Luxembourg and of all cross-border commuters working in the country. This anonymous database, covering the totality of the Greater Region, is available for research purposes. It includes, at the individual level, the municipality of residence, the place of work and certain socio-economic variables. This data is of use to explore the hypotheses discussed above, in particular because they offer a systematic portrait of people recorded at three-year intervals.

However, this administrative record is incomplete in terms of certain key dimensions that are relatively essential to the analysis of residential mobility. This is why the IGSS will be used as a "proxy" to reveal certain characteristics of residential displacements. Our strategy is more orientated towards developing hypotheses than to confirming them. The IGSS database contains, in addition to residential and work locations, variables on nationality, age, marital status, employment status, duration of employment in Luxembourg, salary and the nature of the employment for each of the registered individuals. However, these data are individual and do not offer any information at the household level (income, size or presence of children) or about the level of education or housing tenure status.

Data used

The reference year of 2002 was chosen as the start point for the study of residential mobility. All individuals with their place of residence within the Greater Region were included at the start (n=373,546). This year was chosen in relation to the last census available, from 2001, and the possibility of comparing results with other studies currently under-way. The last available update of the database was used to study possible displacements, which was in 2008. In order to take into account salaries, only salaried persons within the database were then retained (n=251 455); independent workers, while present in the database, do not have a salary assigned to them. The IGSS database contains the annual salary recalculated as a monthly amount. This variable was examined by quartiles, as was the period of professional employment of the individual in Luxembourg. Finally, for reasons of the quality of the data related to the mechanism by which they are updated annually, only salaried employees on the register for at least three months were ultimately included (n=245,392).

The period examined thus runs from 2002 to 2008 and concerns salaried employees living within the Greater Region in 2002. Neither the individuals registered after that date and present in 2008, nor those registered in 2002 but not in the 2008 database were examined. This means that metropolitan movements associated with residential mobility are examined and not those associated with immigration, while metropolitan immigrants resident in the Greater Region are inevitably included in the sample used.

Method

To measure residential movements, the municipality of residence in 2002 was compared with that in 2008. This comparison allowed intra and cross-border displacements to be recorded, as well as the entry into or exit from the country and/or the Luxembourg City agglomeration[3]. It should be noted that residential movements are underestimated, as intra-communal mobility is not taken into account. The place of work of the individual in 2002 was also compared to that of 2008, which allowed the increase or decrease in distance between place of residence and place of work to be recorded (2002-2008)[4]. Finally, the geographical coordinates of the municipality of residence allowed, for those individuals not resident in Luxembourg, the distance to the border in 2002 and 2008 to be calculated, which allowed the analysis to include any movement closer to the border over the period studied.

Three analyses of the binomial logistic regression were constructed in accordance with the working hypotheses discussed above in order to explore, *all else being equal*, the variable cited above. The first analysis attempts, in an initial stage, to show the probabilities of relocation (yes/no) for all salaried employees within the Greater Region. In a second stage, this analysis was repeated with both the salaried employees resident in Luxembourg (n=154,043), and then with salaried cross-border commuters (n=91,349). In a third stage, the two final analyses

3 The borders of the agglomeration of Luxembourg City were determined by combining the morphological analysis of Sohn (2006) and the social indicator work of Carpentier (2007). Thus, the municipalities constituting the urban part of the capital, where continuous built environment is seen, have been added to by the adjoining municipalities with high social indicator figures. The resulting municipalities of the agglomeration are: Bertrange, Contern, Hesperange, Kehlen, Kopstal, Leudelange, Lorentzweiler, Luxembourg-ville, Reckange-sur-Mess, Roeser, Sandweiler, Schuttrange, Steinsel, Strassen and Walferdange.

4 We would like to thank Olivier Klein, who carried out these calculations using a geographic information system (GIS).

were carried out with only those individuals who had changed their municipality of residence in 2008 compared to 2002. The first of these was with residents of Luxembourg (n=38,167), and the second with cross-border commuters (n=22,629). Still using the same illustrative variables, these two analyses attempted to measure the factors influencing the probabilities of moving, in relation to the Luxembourgish border (moving closer, entry and exit) and the agglomeration of the capital city (entry and exit). It should be noted that, given the variables available, the main objective of our analyses was to explore the factors affecting displacements for the cases studied and not the modelling of residential mobility.

Data analysed

Table 6.1 presents the main socio-economic characteristics available in the IGSS' administrative database according to country of residence of the salaried employees within the Greater Region. First, it can be seen that a relatively high proportion of foreign salaried employees reside in Luxembourg, particularly those originating in Europe and countries of the North, at over eight per cent. The three other border countries have relatively similar proportion of nationalities of cross-border salaried employees. The salaried employees residing in the border countries are both younger and to a very large extent more concentrated in the single category (approx. 60% for France and Germany, and 45% for Belgium) than the salaried employees resident in Luxembourg, of whom over 60% are married. These facts suggest life trajectories at the very start of the family and professional cycles often associated with a move/residential establishment.

Table 6.1 – Available data on salaried employees according to country of residence in 2002

	Country of residence in 2002			
(%)	Luxembourg (n=154,043)	France (n=49,195)	Belgium (n=25,349)	Germany (n=16,805)
Nationality				
Luxembourg	56.4	0.9	2	2
France	5.9	92.1	4.7	2.1
Belgium	4.4	1.3	88.6	0.4
Germany	2.2	0.3	0.5	92.7
Portugal	18.9	1.5	0.9	0.3
EU and countries of the North	8.4	3.0	2.6	2.2
Non-EU Europe	2.1	0.0	0.0	0.0
Other	1.6	0.8	0.7	0.4
Age group				
18-24	4.0	10.0	9.0	6.7
25-34	33.1	40.9	41.6	37.9
35-49	46.6	40.4	39.3	47.0
50 and over	16.4	8.7	10.2	8.4

Marital status				
Single	28.7	60.0	46.5	62.9
Married or remarried	62.2	37.2	50.4	35.3
Separated or divorced	8.1	2.5	2.8	1.6
Widowed	1.0	0.3	0.3	0.2
Type of employment				
Non-metropolitan	61.3	67.7	58.2	58.1
Metropolitan	38.7	32.3	41.8	41.9
Duration of employment as of 2002-01-01				
Quartile 1 (recent)	14.3	41.8	39.5	50.7
Quartile 2	21.5	33.3	28.4	28.9
Quartile 3	29.5	16.9	19.8	15.0
Quartile 4 (established)	34.7	8.0	12.4	5.4
Monthly salary in 2002				
Quartile 1 (low)	25.3	29.8	19.0	17.1
Quartile 2	20.4	35.4	28.8	30.8
Quartile 3	24.4	23.5	28.9	29.2
Quartile 4 (high)	29.9	11.3	23.4	23.0

Sources: IGSS 2002, 2008. Authors' calculations.

The proportion of metropolitan-type employment is higher for employees living in Belgium and Germany (42%) than for those living in Luxembourg (38%) and France (32%). Of course, the Luxembourgish residents are those who have been working in the country for the longest, with almost two thirds in the highest two quartiles for duration of employment. On the other hand, around 40% of employees living in France and Belgium are found in the bottom quartile, and 50% of the employees residing in Germany. The employees resident in France are less well positioned in terms of salary (almost 30% are in the bottom quartile for salary, and just 11% in the top quartile, while employees resident in Belgium or Germany (with 19% and 17% respectively in the first quartile, and 23% of both in the top quartile) are in a better position, but behind the employees living in Luxembourg (just 14% in the bottom quartile and almost 30% in the top one). The employees resident in Belgium and Germany are thus less well positioned than those residents in Luxembourg, but better positioned than those residents in France. Access to the Luxembourgish territory tends to be even more influenced by this, which lends support to our hypothesis of borders being approached without being crossed.

Which factors affect the rates of mobility observed?

Table 6.2 presents the results of the first stage of analysis, showing residential movements of employees from the Greater Region according to their country of residence in 2002. A majority of employees from the four countries remains in the same municipality of residence between 2002 and 2008, representing approximately three quarters of people in

Luxembourg, Belgium and Germany, but less than two thirds for the employees resident in France. The proportion of cross-border residential movements out of the total number of persons having moved during the reference period is relatively low, at just 11.7% of moves. The Belgian (23.3%) and German (14.9%) residents experienced more cross-border residential movements than those residents in France and Luxembourg.

Table 6.2 – Residential movements of employees by country of residence in 2002

(%)	Country of residence in 2002				
	Luxembourg[1] (n=154,043)	France[2] (n=49,195)	Belgium[3] (n=25,349)	Gemany[4] (n=16,805)	All (n=216,683)
Residential mobility 2002/2008					
Same municipality of residence	73.5	63.4	75.6	73.9	71.9
Change of municipality	26.5	36.6	24.4	26.1	28.1
Move between 2002/2008					
Borders and residential mobility					
Intra-border movement	88.3	89.3	76.6	85.1	88.3
Cross-border movement	11.7	10.7	23.3	14.9	11.7
Direction of residential mobility					
Exit from Luxembourg	10.6	-	-	-	6.4
Entry into Luxembourg	-	7.9	16.4	10.7	3.9
Within same bordering country	88.3	89.3	76.6	85.1	87.2
Into other bordering country	1.1	2.7	7.4	4.2	2.1

[1] For 10,038 (6.5%) of those resident in Luxembourg in 2002, the individual was no longer present in the IGSS register in 2008.
[2] For 10,873 (22.1%) of those resident in France in 2002, the individual was no longer present in the IGSS register in 2008.
[3] For 4,769 (18.8%) of those resident in Belgium in 2002, the individual was no longer present in the IGSS register in 2008.
[4] For 3,029 (18.0%) of those resident in Germany in 2002, the individual was no longer present in the IGSS register in 2008.

Sources: IGSS 2002, 2008. Authors' calculations.

For the reference period, the balance of entries into/exits from Luxembourg is negative, with exits accounting for 6.4% and entries for 3.9% of the total movements seen between 2002 and 2008. Over 87% of residential movements recorded were within the same country of residence. Of the employees who changed their commune of residence between 2002 and 2008, it is among those employees resident in France (89.3%) and Luxembourg (88.3%) that the highest proportion of movements within the same country of residence are seen. While 7.9% of movements of employees resident in France involve a movement into Luxembourg, employees resident in Belgium (16.4%) and Germany (10.7%) are the ones most likely to move into Luxembourg.

Table 6.3 – Details of salaried employees who changed municipality of residence between 2002 and 2008 by country of residence in 2002

(%)	Country of residence in 2002				
	Luxembourg (n=38,167)	France (n=14,009)	Belgium (n=5,019)	Germany (n=3,601)	All (n=60,796)
Nationality					
Luxembourg	51.3	1.1	2.4	3.6	32.9
France	7.9	93.0	4.6	3.0	26.9
Belgium	6.2	1.5	88.3	0.6	11.6
Germany	2.6	0.3	0.4	89.5	7.0
Portugal	18.8	1.3	0.9	0.5	12.2
EU and countries of the North	8.4	2.0	2.4	2.3	6.1
Non-EU Europe	2.5	0.0	0.1	0.0	1.6
Other	2.3	0.7	0.9	0.4	1.7
Age group					
18-24	8.0	14.5	16.1	9.7	10.3
25-34	50.6	56.7	56.3	53.6	52.7
35-49	33.7	26.5	23.9	33.7	31.2
50 and over	7.7	2.3	3.6	3.0	5.8
Marital status					
Single	42.3	71.9	66.5	72.7	52.9
Married or remarried	47.9	25.7	30.4	24.8	40
Separated or divorced	9.1	2.2	2.9	2.4	6.6
Widowed	0.6	0.2	0.2	0.2	0.5
Type of employment					
Non-metropolitan	40.0	39.1	51.2	50.1	58.7
Metropolitan	60.0	60.9	48.8	49.9	41.3
Duration of employment as of 2002-01-01					
Quartile 1 (recent)	20.8	51.4	56.6	58.2	33.0
Quartile 2	30.7	33.2	26.7	27.9	30.8
Quartile 3	27.4	11.9	12	10.6	21.6
Quartile 4 (established)	21.0	3.5	4.7	3.4	14.6
Monthly salary in 2002					
Quartile 1 (low)	26.8	26.4	18.0	15.1	25.3
Quartile 2	22.9	36.9	29.1	30.0	27.0
Quartile 3	26.2	25.5	31.8	29.9	26.7
Quartile 4 (high)	24.1	11.2	21.1	25.0	20.9
Home-work distance					
Stable 2002-2008	1.9	0.0	0.0	0.1	1.1
Closer 2002-2008	38.7	52.1	55.6	57.0	44.6
Further 2002-2008	59.5	47.9	44.4	42.9	54.3

Sources: IGSS 2002, 2008. Authors' calculations.

Table 6.3 summarises the data available on employees who changed their municipality of residence between 2002 and 2008. No major differences are seen between the nationalities of the employees who moved as opposed to all employees (Table 6.1), except that, to a minor extent, foreign employees are slightly overrepresented in Luxembourg. However, significant differences can be seen in terms of age and marital status. Employees who moved during the reference period tend to be younger and are mostly single, separated or divorced. This corresponds to the classic movements associated with lifecycles, at the start of family and professional careers or at times of rupture to the family situation. This effect is more apparent for employees residing in France and Germany and a little less apparent for those residents in Belgium. This is automatically reflected in the duration of employment in Luxembourg, with over 85% of employees resident in the neighbouring countries in the two most recent quartiles for period of employment within Luxembourg.

Significant differences are seen in terms of employment and employees between countries of residence. A majority (60%) of employees who moved between 2002 and 2008, and resident in Luxembourg and France, have a metropolitan-type job, compared to half (49%) of those resident in Belgium and Germany. Over half of employees who moved residence in France are in the two lower quartiles of salary, while around 60% of those in Belgium and Germany are found in the second and third quartiles. The employees resident in Luxembourg are evenly distributed over the quartiles for salary. For those who moved, it is the employees resident in France who are found at the bottom of the socio-economic scale, while those from the two other neighbouring countries are in an intermediate position. Those from Luxembourg are evenly distributed. The distance between the place of residence and the place of work allows these movements to be put into perspective. Overall, a slight majority of cross-border commuters moved closer to their place of work: around half of the movements documented involved an increase in the distance of journeys to work for the border countries, but almost 60% for Luxembourg residents.

These observations reinforce, for the case of cross-border metropolitan employees, the hypothesis of movement types associated with a move closer to the border and entry into Luxembourg. They show that movements closer to the border tend to involve primarily employees living in France and, to a lesser extent, in Belgium and Germany. The workers resident in these two latter countries have been more stable than those in France, or have been a little more likely to enter into Luxembourg. This supports the two initial hypotheses stating that the employees in a weaker socio-economic position have had less easy access to the Luxembourgish territory. Finally, the movements of employees resident in Luxembourg, even when involving employees with a metropolitan job, appear to indicate an impact of lifecycles. The following section presents three logistic regression analyses of the residential mobility of the same employees, in order to show, *ceteris paribus*, the factors influencing residential stability or mobility (whether intra or cross-border).

Three probabilistic models exploring residential mobility within the Greater Region

The following sections concern, first of all, the factors influencing residential movements within the Greater Region, both in terms of employees resident in Luxembourg and cross-border commuters. In the second and third steps, the factors affecting residential movements are investigated for the same populations, on the basis of their origins/ destination and possible crossings of/movements closer to borders.

Classic factors influencing a residential move

The first analysis (Table 6.4) concerns factors influencing moves of employees working in Luxembourg. Three logistic regression models have been produced, which allow the probability of moving versus that of not moving between 2002 and 2008 to be investigated for: 1) all employees within the Greater Region; 2) employees resident in Luxembourg only; and 3) employees resident in one of the three bordering countries.

At the level of the Greater Region, in comparison to residents of Luxembourg, French employees and those of other nationalities slightly increased their probability of moving, by 1.1 times for the former and 1.3 for the latter. Belgian (0.7), German (0.7) and Portuguese (0.8) employees show a greater tendency towards residential stability. When only those employees resident in Luxembourg in 2002 are considered, the nationalities of the border countries and the other nationalities are, respectively, still in comparison to the Luxembourgish people, 1.2 and 1.4 times more likely to move. In the case of employees resident in a bordering country in 2002, in comparison with those of French nationality, Luxembourgish people (1.6) and those of a non-EU European nationality (2.7) are those most likely to move.

In the three models, the influence of lifecycles is seen to be, as expected, a factor that influences moves. In comparison to employees aged between 35 and 49, those aged 18 to 25 and 25 to 34 are, respectively, almost three and two times more likely to move. Equally, married or remarried employees, and those who are widowed, are considerably less likely to move than single employees, with probabilities of 0.6 and 0.8 respectively of changing their municipality of residence. These two indicators are consistent with the literature, with employees at the start of their careers and single people more likely to move.

Three variables allow us to assess the possible influence of employment on the residential mobility of employees. The type of employment (metropolitan or non-metropolitan) constitutes a significant factor influencing movement of employees at the level of the Greater Region

(1.3), both for cross-border commuters (1.4) and, to a lesser extent, residents of Luxembourg (1.1). Consistent with the variables of age and marital status, logically it can be seen that the duration of employment is longer the lower the probability of movement. In terms of salary, it can be seen from the three models that employees in the highest quartiles for salary are those with the greatest likelihood of moving – regardless of whether they are Luxembourgish residents or cross-border commuters. This partially corroborates the first working hypothesis.

Table 6.4 – Factors influencing residential mobility of employees working in Luxembourg and resident in the Greater Region

	Probability of moving versus not moving between 2002 and 2008		
	Greater Region in 2002 (n=193 274)	Luxembourg residents in 2002 (n=122 709)	Cross-border commuters in 2002 (n=70 565)
Nationality			
Luxembourg	Ref	Ref	1.6***
France	1.1***	1.2***	Ref
Belgium	0.7***	1.2***	0.6***
Germany	0.7***	1.2***	0.6***
Portugal	0.8***	0.8***	ns
EU and countries of the North	Ns	1.1**	0.8***
Non-EU Europe	Ns	Ns	2.7**
Other	1.3***	1.4***	ns
Age group			
18-24	2.7***	2.8***	2.5***
25-34	2.2***	2.0***	2.2***
35-49	Ref	Ref	Ref
50 and over	0.6***	0.6***	0.7***
Marital status			
Single	Ref	Ref	Ref
Married or remarried	0.6***	0.6***	0.6***
Separated or divorced	1.4***	1.3***	1.4***
Widowed	0.8**	0.7***	ns
Type of employment			
Non-metropolitan	Ref	Ref	Ref
Metropolitan	1.3***	1.1***	1.4***
Duration of employment as of 2002-01-01			
Quartile 1 (recent)	Ref	Ref	Ref
Quartile 2	0.7***	0.8***	0.8***
Quartile 3	0.6***	0.6***	0.6***
Quartile 4 (established)	0.5***	0.5***	0.5***
Monthly salary in 2002			
Quartile 1 (low)	Ref	Ref	Ref
Quartile 2	1.1***	1.2***	ns
Quartile 3	1.2***	1.3***	1.1***
Quartile 4 (high)	1.2***	1.2***	1.1***

Home-work distance			
Decrease 2002-2008	Ref	Ref	Ref
Increase 2002-2008	8.0***	10.9***	4.9***
Ref: Reference value ns: not significant Significance level: *<0.10, **<0.05, ***<0.01			

Sources: IGSS 2002, 2008. Authors' calculations.

The change in home-work distance associated with any move is a control variable providing information about the geographical dimension. It is seen that an increase in this distance is likely in the event of a move in all three models. While at the level of the Greater Region a move results in an eight-times greater likelihood of an employee moving further away from their work, it is at the level of Luxembourg that this probability is greatest, with almost 11 times more chance of seeing an increase in this distance. To a lesser extent, moves by a cross-border commuter are also highly likely (4.9) to involve employees moving further away from their place of work.

While increasing urban sprawl is a reality within the Luxembourg employment area, as is reflected in the first three logistic regression models, the above results show that those likely to move are, on the one hand, younger individuals and single persons who are inevitably at the start of their careers, and, on the other hand, those with a relatively well-paid metropolitan-type job. In keeping with the phenomenon of cross-border metropolisation, including that occurring within Luxembourg, this first series of observations supports our working hypotheses, suggesting that those employees best positioned on the social scale are those with the resources necessary to move to the area, and specifically to move to Luxembourg.

What factors affect metropolitan employment for employees resident in Luxembourg?

The second analysis (Table 6.5) concerns intra and cross-border movements associated with the working hypotheses (Map 6.1), here for employees resident in Luxembourg in 2002 only. Those factors influencing a move closer to the capital or an entry into the capital are shown. In addition, those associated with an exit from the agglomeration of the country in 2008 are shown. Four logistic regression models have thus been created with the same variables investigated previously in relation to the factors affecting a move.

The first two models, taking as their reference employees with Luxembourgish nationality, indicate a very low probability that the moves will involve the nationalities of the neighbouring countries or foreign countries, with the exception of "other nationalities". Overall, the mobility options involving a move closer to the agglomeration or an entry into the agglomeration of Luxembourg have (still for employees resident in Luxembourg in 2002) significantly higher probabilities for employees aged over 50 (1.2 for a move closer and 1.3 for an entry), and for individuals with a metropolitan-type job (1.3 for an entry), and for employees in the upper two quartiles for salary (1.2 and 1.3). These results thus support the hypothesis concerning the effects of the metropolisation on the access to and establishment within the vicinity of the capital of those employees who are better positioned on the social scale. They also underline the weak effects of lifecycles on this type of residential trajectory, at least of those linked to the start of a residential career.

Table 6.5 – Factors influencing residential mobility of salaried employees working and residing in Luxembourg

	Probability of employees resident in Luxembourg in 2002... (n=33 705)			
	moving closer to the capital as opposed to another move in 2008	entering the capital as opposed to another move in 2008	exiting the agglomeration as opposed to another move in 2008	exiting Luxembourg as opposed to another move in 2008
Nationality				
Luxembourg	Ref	Ref	Ref	Ref
France	0.6^{***}	0.9^{*}	2.0^{***}	6.3^{***}
Belgium	0.6^{***}	0.8^{***}	1.7^{***}	9.7^{***}
Germany	0.6^{***}	0.6^{***}	1.4^{***}	6.8^{***}
Portugal	0.9^{***}	0.9^{**}	1.5^{***}	0.5^{***}
EU and countries of the North	0.8^{***}	Ns	1.7^{***}	1.4^{***}
Non-EU Europe	Ns	Ns	1.4^{***}	0.2^{***}
Other	0.8^{**}	1.5^{***}	2.9^{***}	0.3^{***}
Age group				
18-24	Ns	0.8^{***}	0.9^{**}	1.2^{**}
25-34	Ns	0.8^{***}	Ns	1.1^{*}
35-49	Ref	Ref	Ref	Ref
50 and over	1.2^{**}	1.3^{***}	1.2^{**}	0.7^{***}
Marital status				
Single	Ref	Ref	Ref	Ref
Married or remarried	0.9^{***}	0.8^{***}	0.9^{**}	Ns
Separated or divorced	Ns	0.9^{*}	0.9^{**}	1.2^{**}
Widowed	Ns	0.4^{**}	Ns	0.3^{**}
Type of employment				
Non-metropolitan	Ref	Ref	Ref	Ref
Metropolitan	Ns	1.3^{***}	1.3^{***}	Ns
Duration of employment as of 2002-01-01				
Quartile 1 (recent)	Ref	Ref	Ref	Ref
Quartile 2	1.1^{*}	Ns	0.8^{***}	0.8^{***}
Quartile 3	Ns	Ns	0.8^{***}	Ns
Quartile 4 (established)	1.1^{**}	Ns	0.7^{***}	Ns

Monthly salary in 2002				
Quartile 1 (low)	Ref	Ref	Ref	Ref
Quartile 2	Ns	Ns	Ns	Ns
Quartile 3	Ns	1.2**	1.1**	Ns
Quartile 4 (high)	Ns	1.3***	1.4***	0.8***
Home-work distance				
Closer between 2002 and 2008	Ref	Ref	Ref	Ref
Further between 2002 and 2008	0.1***	0.1***	5.4***	9.4***
Ref: Reference value ns: not significant Significance level: *<0.10, **<0.05, ***<0.01				

Sources: IGSS 2002, 2008. Authors' calculations.

The two other models used in this analysis show the inverse options: the probabilities of employees resident in Luxembourg in 2002 leaving the agglomeration or the country. These residential exit trajectories involve, to a certain extent, the European nationalities (over 1.4), and the "other nationalities" (2.9). However, the probabilities associated with these exit movements concern especially, as is logical, French, Belgian and German employees, who are, respectively, 6.3, 9.7 and 6.8 times more likely to move abroad than are the Luxembourgish people.

Leaving the agglomeration but remaining in Luxembourg, which is a movement typical of the suburbanisation of the capital, is 1.2 times more likely for an employee over 50 than one between 34 and 49. Equally, employees making such a move are 1.3 times more likely to be an individual with a metropolitan-type job than a non-metropolitan one, and have a 1.4 likelihood of being an employee in the highest quartile by salary than in the lowest quartile. In relation to the probabilities of exiting the country, those most likely to do so are clearly the youngest employees in comparison to those aged 35-49, and those employees least well positioned in the metropolitan labour market.

These two types of trajectories of exiting Luxembourg thus partially support the hypotheses on the more difficult access to and worse positioning within Luxembourg of those employees at the bottom of the social scale. However, while the variables associated with lifecycles certainly affect young employees at the start of their careers and with lower incomes, this influence blurs the possible effect of metropolisation, in particular for those holding a nationality of one of the bordering countries. These trajectories of individuals exiting at the start of their family and professional lives automatically concern the employees who are the least well positioned in terms of employment.

Are cross-border commuters moving closer to the border?

The third analysis (Table 6.6), which includes three models, is also concerned with border dimensions, but this time only for those employees who were cross-border commuters in 2002. The first model investigates the factors influencing a move closer to the Luxembourgish border without moving into the country. The two following models study the factors influencing a move into Luxembourg or into the agglomeration of the capital by 2008. For a better comparison, this latter stage was carried out using the same variables as for the preceding two analyses.

While metropolitan employment appears to increase the residential mobility of cross-border commuters, very few of the variables analysed are found to be significant in this model. It can be seen that, taking the French nationality as the reference, employees of non-EU European nationalities are 6.1 times more likely to move closer to the border, followed by employees of Luxembourgish nationality (2.9) and "other nationalities" (ranging from 1.3 to 1.5). As in the preceding models relating to relocations and residential movements of employees resident within Luxembourg, the employees in the highest quartile by salary are slightly more likely to move closer to the border than their counterparts in the lowest quartile. This means that, overall, neither lifecycle effects nor metropolisation affect the probability of a move closer to the border in relation to another type of residential movement.

The last two models examine the probabilities of cross-border commuters moving into Luxembourg or the agglomeration of its capital as opposed to making another type of move; this is the area we consider to be "for the privileged" in our initial premises. These multiple variables support our working hypotheses and present highly significant degrees of correlation, in particular those associated with the international character of Luxembourg and the classes of employees at the top of the social scale: the transnational classes.

Also supporting our working hypotheses, it can be seen that a move into the country of Luxembourg is – in relation to those aged 35 to 49 – 1.6 times likely for employees aged over 50, and a move into the agglomeration is twice as likely for this same category of employees. Only separated or divorced individuals are more likely to make the same kind of move than single persons. The two variables associated with metropolisation are seen to be statistically significant and also support our working hypotheses. Employees with metropolitan-type employment and belonging to the highest quartile for salary are, respectively, 1.3 and 1.4 times more likely to move into Luxembourg than non-metropolitan workers and those in the lowest quartile by salary. For a move into the agglomeration, these same probabilities rise to two in the case of employees in metropolitan employment and 1.7 for those in the highest quartile by salary. These observations appear to reinforce the influence of metropolisation on access to the country and the area of its capital for cross-border commuters, while clearly minimising, in this case, the impact of lifecycle effects.

Nationality certainly appears to play a role in the likelihood of entering or exiting Luxembourg. Taking as the reference French employees, who are less well positioned on the social scale, it can be seen that employees of Luxembourgish (12.6), Portuguese (7.1), or non-EU European (6.4) nationality are, with different degrees of statistical significance, considerably more likely to move into Luxembourg. Still in relation to those of French nationality, Belgians (1.8) and those belonging to an EU country or a country of the North (2.8) are also more likely to enter the country, but Germans are less likely to do so. These probabilities are not as marked for the case of a move into the agglomeration, with the exception of European employees (2.1), and those from non-European countries and those of the North (6.1), and for those of "other nationalities" (2.8). This shows the relative attractiveness of the Luxembourgish metropolitan area to the nationalities associated with the highly-skilled classes of workers.

Table 6.6 – Factors influencing the residential mobility of salaried cross-border commuters working in Luxembourg

	Probability of employees resident in a bordering country in 2002 of... (n=22 081)		
	moving closer to the capital as opposed to another move in 2008	entering the capital as opposed to another move in 2008	exiting the agglomeration as opposed to another move in 2008
Nationality			
Luxembourg	2.9***	12.6***	2.4***
France	Ref	Ref	Ref
Belgium	1.4**	1.8***	1.2**
Germany	Ns	0.8**	0.5***
Portugal	1.5**	7.1***	2.0**
EU and countries of the North	1.3**	2.8**	2.1***
Non-EU Europe	6.1**	6.4**	6.1**
Other	Ns	2.1***	2.8***
Age group			
18-24	Ns	0.8*	Ns
25-34	Ns	0.8***	Ns
35-49	Ref	Ref	Ref
50 and over	0.8**	1.6***	2.0***
Marital status			
Single	Ref	Ref	Ref
Married or remarried	Ns	Ns	Ns
Separated or divorced	Ns	1.6***	1.5*
Widowed	Ns	Ns	Ns
Type of employment			
Non-metropolitan	Ref	Ref	Ref
Metropolitan	Ns	1.3***	2.0***

Duration of employment as of 2002-01-01			
Quartile 1 (recent)	Ref	Ref	Ref
Quartile 2	Ns	Ns	Ns
Quartile 3	0.8^{***}	Ns	0.8^{**}
Quartile 4 (established)	0.8^{***}	Ns	Ns
Monthly salary in 2002			
Quartile 1 (low)	Ref	Ref	Ref
Quartile 2	Ns	0.8^{***}	0.8^{**}
Quartile 3	Ns	Ns	Ns
Quartile 4 (high)	1.1^{**}	1.4^{***}	1.7^{***}
Home-work distance			
Closer between 2002 and 2008	Ref	Ref	Ref
Further between 2002 and 2008	0.0^{***}	0.1^{***}	0.0^{***}
Ref: Reference value ns: not significant Significance level: *<0.10, **<0.05, ***<0.01			

Sources: IGSS 2002, 2008. Authors' calculations.

Discussion/Conclusion

This chapter has investigated, on the basis of administrative data, the influence of certain factors on residential mobility within the metropolitan region of Luxembourg. On the basis of four types of residential trajectory, the residential displacements of employees have been analysed: moves closer to the Luxembourgish border; crossing the border to move into the country/capital; exiting to a bordering country; or a move not involving the Luxembourgish border. For the reference period of 2002-2008, few cross-border moves are seen, with the majority of moves taking places within the same country of residence. While the likelihood of moving into the Greater Region appears to be influenced by lifecycles (age, marital status), the process of metropolisation, especially the entry into Luxembourg or the capital, appears to be related to the effects of metropolisation (type of employment, salary). On the one hand, the nationality of employees moving into Luxembourg and their type of employment appear to show a polarisation of "the privileged." These findings may give the Luxembourgish case a special status in connection to theories and concepts related to socio-spatial segregation that are usually applied at scales linked to cities and neighbourhoods. Indeed, the limited size of the country and the proximity of the borders to the capital give the observed segregative dynamics a (trans)national dimension. On the other hand, however, while our findings show that trajectories of exit from the country involve less well-positioned employees, these also involve individuals at the start of their residential and professional careers, which complicates the factors influencing an exit. In this context, it is also difficult to clearly interpret this part of our results in the specific context of the processes of segregation and metropolisation. However, this exploration of residential movements within Luxembourg has allowed the working hypotheses to be refined and new hypotheses to be developed.

Residential moves for those who can afford it?

High salaries, which are associated with metropolitan jobs, are clearly linked to a likelihood of moving to the Luxembourgish employment area. Access to Luxembourg and its capital thus appears to be primarily granted to those at the top of the social scale. Both the nationalities in question and the job and salary characteristics reinforce the hypothesis that movement into the country is associated with those high on the social scale and the transnational classes of workers. It is, however, more difficult to reach conclusions about the dynamics at work at the bottom of the scale, as higher salaries are also associated with exit trajectories. Two categories of worker can thus be discerned among those moving out of Luxembourg: a disadvantaged class without the means to live within Luxembourg, and another better-off class that participates in the dynamic of cross-border peri-urbanisation; the latter individuals take advantage of cross-border differentials in property prices to ensure that they are able to become dwelling owners. It will thus be useful to investigate in more detail nationalities and changes in housing tenure status, and perhaps help to identify and make a distinction between specific phenomena related to the cross-border region of Luxembourg and processes merely related to urban development, urban sprawl and access to home ownership.

From residential movements to segregative dynamics?

Does the emerging Luxembourgish metropolis, however, constitute a true transnational social space? While we have not been able to confirm our working hypotheses in this regard, we have also not been able to completely refute them. The findings show that the process of exclusion appears to be of little relevance at the level of Luxembourg and the border regions. On the contrary, social withdrawal of the privileged classes appears to emerge from the trajectories observed, as we illustrated with data (Lord and Gerber 2009). Repeating the analyses here with more detailed data at the level of the Luxembourg City agglomeration would allow this hypothesis to be examined more closely. The process of the development of exclusive districts of foreign workers with their own networks appears to apply here. The inclusion of self-employed workers and officials of European Institutions, as well as the study of more varied residential movements (temporary moves, dual residency, second dwelling, etc.) should also be considered in the examination of cross-border residential mobility. The inclusion of housing tenure status and information regarding dwelling ownership would also be crucial in this context. Do people enter Luxembourg to become dwelling owners or simply to work temporarily, independently of their housing tenure status? And, conversely, do they leave Luxembourg or its agglomeration to buy property, or solely to reduce their rental costs? Some answers to questions could be provided by a study of the workers who have left Luxembourg to live in neighbouring countries while continuing to work in Luxembourg (Carpentier 2010).

What tensions are residential trajectories subject to?

Residential trajectories of individuals and households are subject to a complex collection of forces. While this chapter has been able to gain a more detailed understanding of the reality concerning the factors influencing residential displacements, it is nonetheless some way from explaining the logic that underlies the residential mobilities of individuals and their households. While the longitudinal data analysed have been able to show the effects of lifecycles and metropolitan employment on residential movements, this mobility is certainly also subject to factors related to opportunities and attachment to the local level (Enaux and Gerber 2008). Several factors are likely to interact, particularly family, professional and daily mobility factors. A move closer to or further away from the border is thus likely to be a response to multiple tensions that are traded off by households on a daily basis. It is at the household level and with the analysis of qualitative aspects (daily mobility, employment, residential preferences) that a more complex investigation could be carried out.

Note

This article benefited from co-funding from Fonds National de la Recherche (Luxembourg). It was also presented at the *Coloquio Fronteras Contemporaneas: Miradas Desde Las Ciencias Sociales* in Tijuana in 2009 by S. Lord, P. Gerber and C. Sohn.

CHAPTER 7

Metropolitan Regions and Cross-border Demographic Integration

Olivier WALTHER

Far from resulting in the general dispersal of the population, the growth and hyper-fluidity of the exchange of goods, persons and capital are contributing to a renaissance of those metropolitan regions that are best connected at the global level (Taylor, Walker and Beaverstock 2002, Taylor 2004, Hall and Pain 2006, Rozenblat and Pumain 2007). The development of means of overcoming distance are allowing increased accessibility and connectivity between the main hubs of economic and technological development, resulting in increased interregional and international competition for access to the resources offered by central locations. In this process, metropolitan regions tend to exert a strong attractive force over not just creative and highly qualified individuals representing the knowledge economy, who find there a specialised and attractive employment market (Florida 2005), but also over unskilled individuals whose economic activity supports the development of an overlying service economy (Sassen 2001).

As a result, metropolitan regions are experiencing demographic growth, which indicates the intensity of human flows on multiple levels. As Europe's rate of natural increase is low, the demographic growth of metropolitan regions results from population movements within European countries and from international migration. In Germany, for example, the relocation of inhabitants of the states of former East Germany contributes to the concentration of the German population in the northern and southern German metropolitan regions (Hamburg and Munich respectively), (Bade 2003). A similar observation can be made in France, whose population remains heavily concentrated in the Greater Paris region and certain other major French urban agglomerations, in particular those of the south (Bordeaux, Toulouse, Montpellier, Marseille, Nice), whose appeal to migrants is well established (Laganier and Vienne 2009). These national contributory factors are added to by international migration, which benefits for example the London region (UK National Statistics 2007, Hatton and Tani 2005) and the metropolitan regions of Belgium, the Netherlands, Luxembourg and Switzerland. In this latter country, the highest rate of demographic growth in both absolute and relative terms over the past 30 years has been seen in the country's two major metropolitan regions, Zurich and the Lake Geneva region (Da Cunha and Both 2004, Schuler *et al.* 2007).

Urban growth is not, however, restricted solely to growth in the major centres at the global level. It also contributes to the renewal of the forms and scales of urban spaces at the intra-metropolitan level. The demographic evolution of the metropolitan regions has thus been strongly influenced by the processes of suburbanisation, which has been taking place since the 1960s (for a review of the literature in Europe, see Guillain, Le Gallo and Boiteux-Orain 2006). From this perspective, the development model based on car usage and individual housing remains dominant (Carpentier 2007, Lord and Gerber 2009), although some authors have provided evidence of the process of gentrification, which benefits above all the new middle class attracted by the social and spatial characteristics of central areas (Ley 1996, Hamnett 2003).

The cross-border metropolitan areas are clearly affected by this double phenomenon of concentration at the global level and loosening at the intra-metropolitan level. Examining the cases of Luxembourg, Basel and Geneva, this chapter's primary objective is to document the effects that national boundaries may have on demographic dynamics in a metropolitan context. Does urban growth lead to a reduction in demographic differentiation, or does it, on the contrary, accentuate disparities, resulting in the increasing fragmentation of infra-metropolitan regions located on either side of national frontiers?

The chapter is arranged in four parts. The first part deals with spatial integration and presents the two major contradictory trends that are affecting cross-border residential integration: the growth in the linkages uniting the European nations and the fragmentation increasingly affecting infra-metropolitan areas. The second part presents the case study and the methodology used, which combines the analysis of transects and indicators of spatial autocorrelation. The third part provides a descriptive analysis of dynamics of cross-border regions, stretching from Luxembourg to Geneva. The fourth part presents the results of the analysis of spatial autocorrelation for the metropolitan areas of Luxembourg, Basel and Geneva, in order to show groups of municipalities with similar demographic profiles and the possible effects of borders. The final part concludes by stressing the unique case of Luxembourg in terms of cross-border demographic integration.

Cross-border metropolitan integration

Multiple definitions of the term "spatial integration" have been given, particularly in the content of studies relating to European structure at the level of nation states or cross-border regions (Anderson and Wever 2003, Brenner 2004, Dabinett and Richardson 2005, Hansen and Serin 2007). In the most popular meaning, spatial integration is understood to be a process that allows the convergence of different spatial systems by means of the interactions between their members. Integration thus reflects "the creation and maintenance of intense and diverse patterns of interaction and control between formerly separate social spaces" (Lee 2009: 387). In their analysis carried out within the framework of the *Study Programme on European Spatial Planning*, De Boe, Grasland and Healy (1999) argue that spatial integration should be considered as a process leading to a reduction in the structural differentials within a system and greater homogeneity.

Whether or not this dimension should be taken into account continues to be a subject of discussion in the literature, in particular because a process of homogenisation need not necessarily imply that flows are being exchanged between cross-border regions; homogenisation can result from the internal dynamics of the regions in question. Conversely, it is often noted that highly integrated areas, in terms of professional interactions, for example, are highly heterogeneous, with their integration resulting precisely from the existence of the spatial differentials. The relations between interactions and homogeneity are thus far from being automatic.

Demonstrating the value of also considering integration from the point of view of homogenisation, Grasland, De Boe and Healy (1999) look at the GNP of 22 Italian provinces from 1951 to 1991, showing both the process of territorial homogenisation (reduction of inequalities in each region examined) and the process of spatial diffusion with barrier effects for the periods examined. The authors note that such analyses could be conducted using other variables than GNP, which do not directly measure relations between the areas, such as variables relating to education, living conditions and demography – these would have the advantage of being more readily available than the flow variables. This approach will be used in this work, taking into account the two contradictory trends affecting the process of cross-border integration in Europe.

On the one hand, it is clear that the intensification of global trade and the harmonisation of the EU's different economic, legislative and financial systems have led to profound changes in the traditional functions of European national boundaries. The contact and exchange functions of border regions have thus progressively become more important than the barrier and control functions traditionally assigned to national frontiers (Anderson 1996, O'Dowd 2002). This evolution has led to the growing functional integration of cross-border metropolitan regions in terms of economic exchanges, transport networks, labour markets and property markets (Ratti 1994, Reitel 2007). In the residential domain, too, the opening up of European frontiers has allowed an increasing number of individuals to establish themselves on the side of the border that they deem to be most advantageous. In Geneva, Luxembourg, Nijmegen and Maastricht, for example, certain workers choose to commute into their own country from outside, generally in order to access more affordable real estate (INSEE-OCSTAT 2008, Carpentier 2010, Van Houtum and Gielis 2006). In addition to this, institutional dynamics result in a growing number of metropolitan regions implementing governance projects (Perkmann 2003, 2007) that aim to respond to metropolitan development issues. By favouring cooperation between actors located on different sides of frontiers, these initiatives can, for example, contribute to better coordination of the provision of transport networks, respond to a certain extent to imbalances in the labour and accommodation markets to avoid too great a level of functional specialisation, simplify administrative processes or develop joint development strategies.

On the other hand, it is clear that the construction of metropolitan territories takes place on the basis of increasingly fragmented areas (Veltz 1996). In Western Europe, as in North America, the general loosening of urban fabric has resulted in the construction of vast urban agglomerations that are discontinuous in residential terms. These urban agglomerations are subject to spatial reorganisation resulting from both the economic capacity of households – notably for reasons of access to real estate – and individual aspirations, which leads to the segregation rather than the homogenisation of residential space (Musterd 2005). In the case of cross-border metropolitan regions, this phenomenon can be amplified by national differentials. The cross-border cities thus constitute a special case, the interest of which is that they combine sometimes very different population dynamics and experience unique imbalances linked to countries' histories. In Geneva and Basel, for example, the historic stability of frontiers has led to the strong differentiation of national territories from the point of view of population density between Switzerland and its French and German neighbours (Schuler *et al.* 2007). Elsewhere in Europe, national dynamics have strongly affected border regions that served as a military glacis (Foucher 1991) or were defended by strongholds such as Dunkirk, Lille and Valenciennes in France, or Gand, Tournai and Namur in Belgium (Denys 2002). This historical heritage can exercise a strong degree of inertia over the process of demographic integration.

Case study and methodology

Luxembourg, Basel and Geneva

The decision to study Luxembourg, Basel and Geneva is based on a number of factors. Firstly, the functional metropolitan areas of these cities have comparable population levels: 930,000 inhabitants for Luxembourg; 950,000 for Basel; and 807,000 for Geneva in 2006 (ESPOM 2010). The populations have a significant cross-border component, which is, however, higher in Luxembourg (44.9%) than in Basel (39.8%) or Geneva (34.2%), (Sohn, Reitel and Walther 2009). Secondly, these three cases have a roughly similar functional structure, marked by a high level of focusing of the peripheral areas on the main centre of employment. Finally, Luxembourg, Basel and Geneva are cities whose level of influence at the European level is significantly higher than would be suggested by their population level (Rozenblat and Cicille 2003). This is explained by heavy specialisation in certain knowledge-intensive services (Luxembourg and Geneva) and industries (Basel), and by the presence of numerous international agencies.

The analysis of demographic dynamics is conducted at various levels. In the initial descriptive section, all municipalities situated on the border running from north of Luxembourg to south of Geneva are examined. In a more analytical section, the study is restricted to the metropolitan areas of Luxembourg, Basel and Geneva. The cross-border metropolitan areas are identified on the basis of the work of Sohn and Walther (2009) for Luxembourg and of Dessemontet, Perlik and Schuler (2005) for Basel and Geneva. They include all municipalities for which the proportion of commuters working in the central agglomerations was higher than 16% of the economically active population in 2002 (Luxembourg) or 2004 (Basel and Geneva). The population data at the municipality level was provided by the national statistical services of Luxembourg, Switzerland, France and Belgium (STATEC, OFS, INSEE, INS) and of the various German states (Saarland, Rhineland-Palatinate, Baden-Württemberg). An effort to achieve harmonisation led to use of the years 1960, 1980 and 2000, or the closest years available.

Methods

The analysis of demographic differentials was based on two methods: transects and spatial autocorrelation. The analysis of transects is a suitable method to demonstrate cross-border differentials. Taking the centres of the metropolitan regions of Luxembourg, Basel and Geneva as the origins, these transects allow analysis of population development in several geographical orientations from 1960 to 2000. Given that population densities decrease in a regular manner between the metropolitan centre and the periphery (Newling 1966), several ideal forms can be distinguished.

Figure 7.1 – Density and distance from metropolitan centre

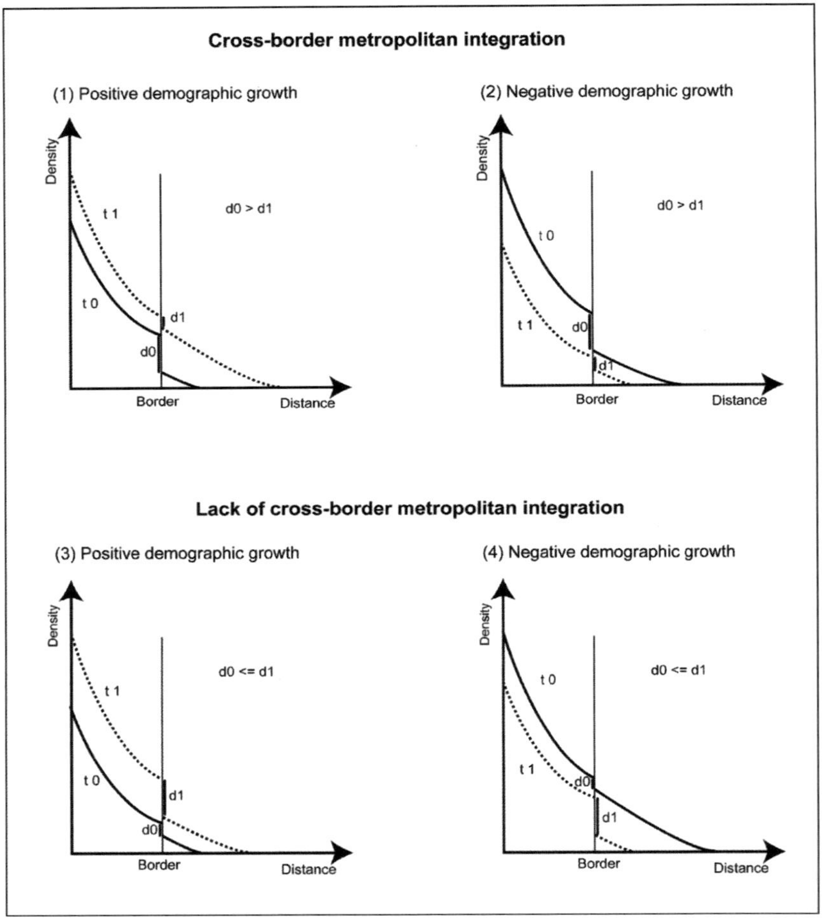

Source: Walther.

Figure 7.1 shows in this respect that demographic integration is seen in areas in which the differential observed at t0 (d0) is higher than that observed at t1 (d1). Integration thus marks a reduction in the disparities observed along the transect on both sides of the national frontier. This property is valid for both positive (1) and negative (2) growth. Absence of demographic integration, in contrast, characterises those areas in which the differential observed at t0 (d0) is equal to or less than the differential observed at t1 (d1). As before, this property is valid whether concerning positive (3) or negative (4) growth.

These theoretical options do not exhaust the possible configurations, as the example of Luxembourg, Basel and Geneva show in the remainder of the present study. It is in reality rare for the demographic differential to be able to be measured exactly at the frontier. It is thus important to extend the observation of differentials to a band of variable width, in accordance with the dimensions of the metropolitan region in question. In addition, demographic developments can be heavily influenced by national policies, with the result that municipalities experience positive demographic growth on one side of a frontier and negative growth on the other.

The transects cannot be carried out on the basis of the existing municipality boundaries, as these spatial units have highly divergent sizes. In order to attenuate these variations, municipality values are combined with the method of inverse distance weighting, on the basis of the premise that each unit is more affected by the neighbouring municipalities than by municipalities that are further away, within a defined research radius.

The analysis of the transects is complemented by an analysis based on spatial autocorrelation methods carried out with the GeoDa software (Anselin, Syabri and Kho 2006). Spatial autocorrelation is based on Tobler's law, which postulates that, all else being equal, neighbouring spatial units are more likely to have similar values than units that are further apart. The main purpose of the spatial autocorrelation indices is to test the degree of spatial dependence between neighbouring areas (see Baumont, Ertur and Le Gallo 2004 for an application of this to France). In this study, Moran's indices are used to take account of the evolutionary dynamic of the population. Firstly, the Moran's I index is used to calculate the relation between the covariance of the contiguous sites and the global variance of the characteristic in question (Zaninetti 2005). It determines the relationship between the values of a unit and the values of the neighbours of this unit: an index higher than 0 indicates positive spatial autocorrelation, i.e. that the neighbouring sites are more similar to it than are distant sites, while an index of less than 0 indicates negative spatial autocorrelation, showing that distant sites are more similar to each other than neighbouring sites.

As the Moran's I index provides just a single value for each metropolitan region, it is necessary to then calculate the local Moran's index (Anselin 1995). This local indicator of spatial association is the vectorial product obtained between the neighbours of a unit and the unit itself, for each municipality in the zone being studied. It shows the degree of similarity or dissimilarity between a municipality and its neighbours and compares the value with that which would be observed if the distribution of the population were random. It allows clustering phenomena to be demonstrated for a particular spatial unit, and the demonstration of the presence of spatial units with extreme values. A standardised score is assigned to each unit in the zone being studied: a positive and high z-score indicates that the municipality is surrounded by similar positive or negative values, while a negative z-score indicates that the municipality is surrounded by dissimilar neighbours. The cluster gradient thus moves from high negative values to high positive values.

These results can be represented in the form of a graph with four quadrants, showing on the x axis the standardised population density of each municipality and on the y axis the spatial lag corresponding to the average standardised value of the neighbouring spatial units (Anselin 1996). Four types of local spatial association between a municipality and its neighbours are identified: HH (high-high), HL (high-low), LH (low-high) and LL (low-low), with HH corresponding to high density municipalities surrounded by spatial units with a similar profile for example. In this study, only results significantly different from 0, at a 5% significance level, are classed as spatial autocorrelation and thus shown cartographically.

Growth and population densities of border regions

In Western Europe, the border regions have highly heterogeneous demographic profiles. By combining annual population growth between 1980 and 2000 with the municipality density in 2000, four distinct profiles can be identified at the level of the border regions from Luxembourg to Geneva (Map 7.1).

The first profile primarily concerns the main urban centres studied (Luxembourg, Strasbourg, Basel, Zurich, the Lake Geneva region, Annecy, Lyon and Chambéry). These centres show both strong demographic growth, defined as an annual rate of growth of over 2.0%, and high density: above 130 inhabitants per km^2. In Luxembourg, the area forms a discontinuous belt around the south of the capital. In Geneva, the municipalities in question are located around the edge of Lake Geneva and in Pays de Gex, to the west of the city. In Basel, the territorial units conforming to this profile are located in a radius of 15 km to the south of the city.

The second profile concerns those municipalities with a typically peri-urban demographic profile: an annual rate of growth of over 2.0% and a density below 100 inhabitants per km^2. In Luxembourg, the relevant territorial units are located to the north of the capital and along the French border. Two types of municipalities located at the periphery of the Geneva metropolitan region are represented in particular: the one located in the Swiss canton of Vaud to the north east – which benefits from the joint dynamism of Geneva and Lausanne and the significant shortage of accommodation in the Lake Geneva region – and the one located to the south and west of Geneva in France, which is very popular with cross-border workers. This also applies to the Alpine municipalities that focus on tourism or specialist industries in Savoie and Haute-Savoie. Basel has few territorial units of this type.

The third type of profile applies to municipalities of two very different types: city centres and former industrial and/or mining areas. In these municipalities, the average annual rate of growth is generally zero or negative, while the density is higher than 130 inhabitants per km^2. The city centres of Basel, Geneva and Luxembourg also have this characteristic[1]. Between 1960 and 2000, Basel experienced the least favourable annual demographic change (-0.4%), followed by Luxembourg (-0.2%) and Geneva (+0.6%). While the two Swiss cities have seen significant growth in their populations within the urban agglomerations, Luxembourg is atypical in that the average annual rate of demographic growth of its agglomeration has remained weak (+0.5% from 1981 to 2000). This feature can be explained by the fact that the transformation of the Franco-Luxembourg cross-border industrial area from the 1970s onwards did not lead the French, Belgian or German populations to move towards the emerging financial centre of Luxembourg but rather led to a spectacular growth in cross-border commuting, which became structural rather than temporary. Thus, although the structure and distribution of employment was completely transformed in the region, the population structure has only evolved slowly and adapted by means of recourse to increased transportation (Carpentier and Gerber 2009). The former mining and steel-working areas of Saarbrücken, Liège and Metz-Thionville have also seen zero or negative economic growth. New arrivals linked to the growth of cross-border commuting to Luxembourg have not been sufficient to compensate for the demographic decline in these municipalities. The barrier effect is especially pronounced between the south of Luxembourg and the French neighbours in the region from Esch-sur-Alzette to Longwy, and between the central Basel municipalities and their French and German neighbours.

1 The period considered does not, however, take into account the demographic revival seen in the city centres, a change that began in the 2000s. Like other European cities, Luxembourg saw a first peak in its population at the end of the Fordist industrial era (1979: 79,596 inhabitants), followed by a 20-year period of decline. The city is experiencing once again sustained growth in its population with 94,034 inhabitants in 2011.

Map 7.1 – Cross-border demographic dynamics, 1980-2000 (see colour section)

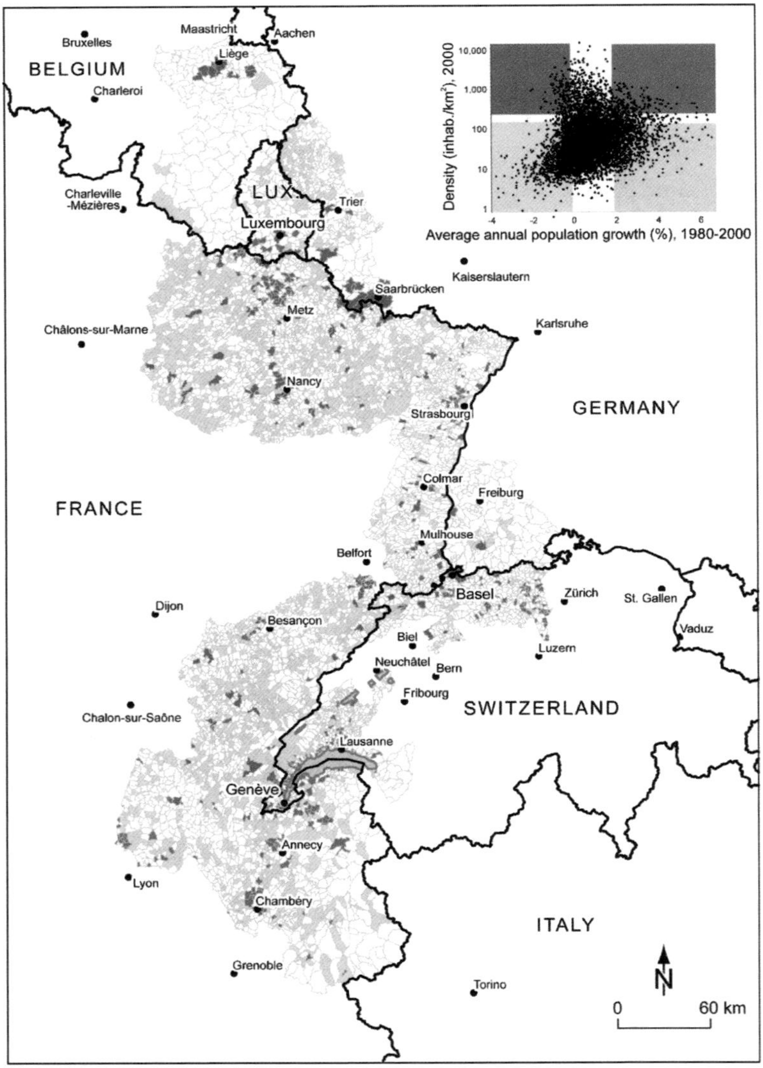

Source: National statistical offices. Cartography: Walther.

The fourth profile is characterised by weak demographic growth (>=0.0%/year) and low density (<100 inhab./km^2). It applies to the rural municipalities with little industry in the Belgian Ardennes, Meuse, the Eifel, the Vosges and the French Jura. The demographic decline here is especially pronounced due to the joint impact of the fall in economic activity and the ageing of the population. A deep divide is seen between the whole of Luxembourg, which has one of the highest rates of demographic growth in the EU, and the German Eifel.

The specific features of the Luxembourg situation are shown by the transects carried out on a selection of profiles in each of the metropolitan regions. Moving towards Arlon in Belgium (and Trier in Germany, not shown), the densities do not increase by a similar amount on each side of the border, which cannot be characterised as a process of integration (Figure 7.2).

Figure 7.2 – Densities between Luxembourg and Arlon, and Luxembourg and Lorraine, 1960, 1980, 2000

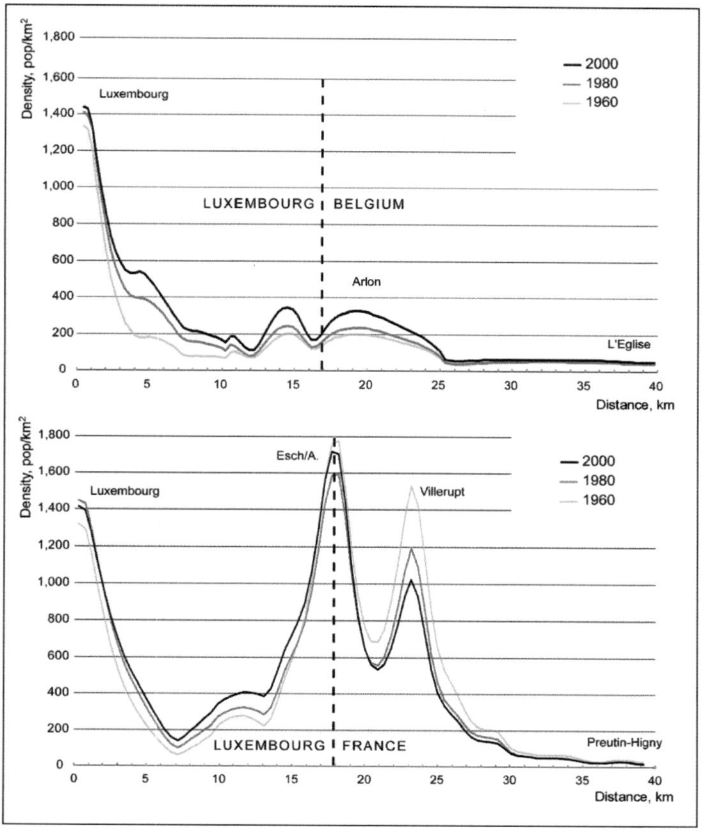

Source: National statistical offices. Calculations: Walther and Dickes.

Figure 7.3 – Densities between Geneva and Pays de Gex, and Basel and Jettingen, 1960, 1980, 2000

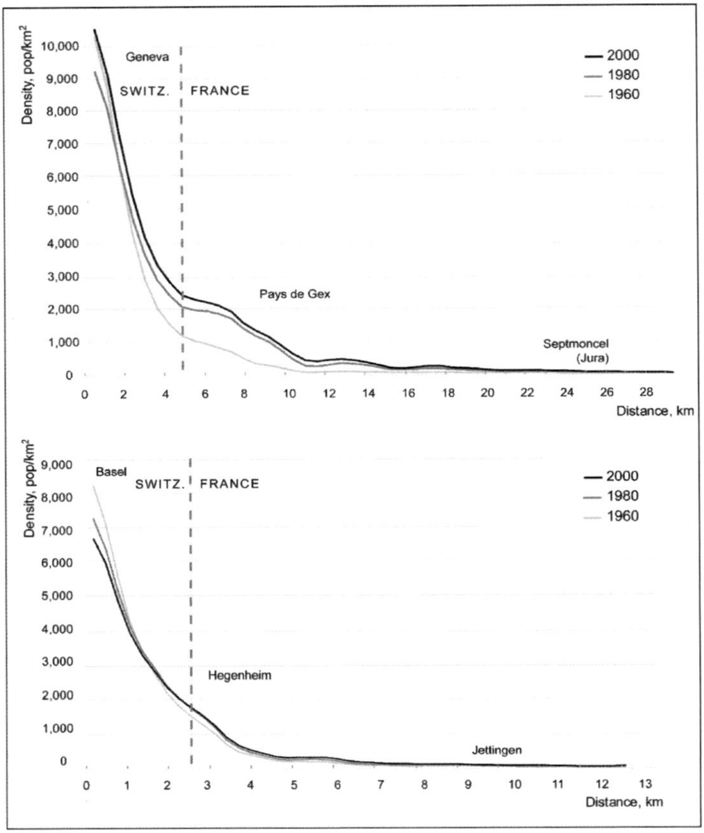

Source: National statistical offices. Calculations: Walther and Dickes.

The transects carried out in the former Franco-Luxembourg steel-working region in the direction of Lorraine (Preutin-Higny) show a significant contrast between the Luxembourg municipalities, with growing densities, and the Lorraine municipalities, where densities fell over the period in question. The presence of an especially urbanised zone to the south of Luxembourg complicated somewhat the interpretation of the transect, which deviates from the theoretical pattern shown in Figure 7.1, in which densities decline as distance from the centre increases. However, this special case illustrates the fact that industrial decline has been more severe in France than in Luxembourg.

The transect carried out between Geneva and Pays de Gex in France is undoubtedly the one that shows most clearly a process of cross-border demographic integration (case 1 in Figure 7.1). The densities of the French Pays de Gex have grown more significantly than those in the canton of Geneva, which has contributed to reducing the differential between these two countries. The situation is less marked between Basel and Jettingen in Alsace, due to the comparatively weak growth of the population in this metropolitan region (Figure 7.3).

Strong cross-border demographic disparities

In this section the cross-border differentials are analysed using spatial autocorrelation methods. These methods allow the groups of municipalities with similar average annual demographic growth over the periods examined (1960-1980, 1980-2000, 2000-2008) to be identified, and the effect of national borders on the demographic dynamics of the metropolitan regions of Luxembourg, Basel and Geneva to be assessed.

Table 7.1 presents the results of the Moran's I index by degree of contiguity of the municipalities and the significance of the results ($p<0.001$). It allows an estimate to be made of the average degree to which a municipality exerts a spatial influence over its neighbours.

Table 7.1 – Moran's I values, 1960-2000

	Luxembourg		**Basel**		**Geneva**	
Degree	*Moran's I*	*p-value*	*Moran's I*	*p-value*	*Moran's I*	*p-value*
1	0.464	0.001	0.354	0.001	0.344	0.001
2	0.336	0.001	0.203	0.001	0.245	0.001
3	0.197	0.001	0.090	0.001	0.159	0.001
4	0.032	0.032	-0.017	0.216	0.132	0.001
5	-0.072	0.001	-0.021	0.114	0.095	0.001
6	-0.165	0.001	-0.014	0.233	0.020	0.109
7	-0.175	0.001	-0.010	0.279	0.020	0.099
8	-0.153	0.001	-0.015	0.203	-0.064	0.004

Calculations: Walther and Dickes. Note: contiguity weight matrix is used (queen's case).

A positive spatial autocorrelation can be observed between the neighbouring municipalities in the three metropolitan areas, but this is stronger in Luxembourg (0.46) than in Basel (0.35) or Geneva (0.34). In order to check that these results are not due to a purely spatial effect dependent on the size of and form of the municipalities, the calculation included more distance degrees of proximity. It would be expected that the spatial autocorrelation would diminish with the number of degrees.

For Luxembourg and Basel, a positive spatial autocorrelation can also be seen up to the third degree of proximity. A negative autocorrelation can then be observed for the subsequent degrees, but only the values for Luxembourg are significant. In Geneva, a positive and significant spatial autocorrelation can be seen up to the fifth degree of proximity; the following ambiguous or negative values are not significant. These findings confirm that the rates of growth seen between 1960 and 2000 conform well to an overall pattern with a similar degree of contiguity for Luxembourg and Basel and a slightly higher one for Geneva.

It should also be noted that the spatial autocorrelation was stronger for the period from 1960 to 1980 than for that from 1980 to 2000 for all metropolitan areas. This is particularly the case for Geneva, where the Moran's I index went from 0.318 to 0.138, and for Basel, where it fell from 0.329 to 0.227 between the two periods examined (Table 7.2).

Table 7.2 – Moran's I values, 1960-1980, 1980-2000, and 1960-2000

Metropolitan areas	1960-1980	1980-2000	1960-2000
Luxembourg	0.393	0.386	0.464
Basel	0.329	0.227	0.354
Geneva	0.318	0.138	0.344

Calculations: Walther and Dickes. Note: contiguity weight matrix is used.

This development indicates that the municipalities within the metropolitan area have increasingly heterogeneous rates of demographic growth. This could be interpreted in terms of the timing of the general trend of suburbanisation experienced by all metropolitan regions in Europe. The process of suburbanisation, which began in the 1960s, gradually extended to areas increasingly far away from the city centres. However, in the 1980s, a trend can be observed of a shrinking of the disparities between the profiles of those municipalities close to the centre and the suburbs, linked to the intensive use of cars, the exhaustion of land of construction, and the reduction of the size of households following the ageing of suburban zones. The highest levels of growth were then seen at long distances from the centre, while a certain demographic revival is seen in the city centres, in favour of national and international migrants and, secondarily, of the installation of certain demographic groups working in the creative or knowledge-based industries.

This general demographic development at the intra-metropolitan level is illustrated by the analysis of a transect between Luxemburg and Trier (Figure 7.4). This graphic shows the temporal interval between the two average annual rates of growth for the period 1960-1980, which essentially concerned the municipalities close to Luxembourg City, and the rates of growth for the period 1980-2000, which concerned the peri-urban area. The growth of the German municipalities was influenced both by the suburbanisation of Trier and the cross-border peri-urbanisation of Luxembourg.

Figure 7.4 – Rates of demographic growth between Luxembourg and Trier, 1960-1980 and 1980-2000

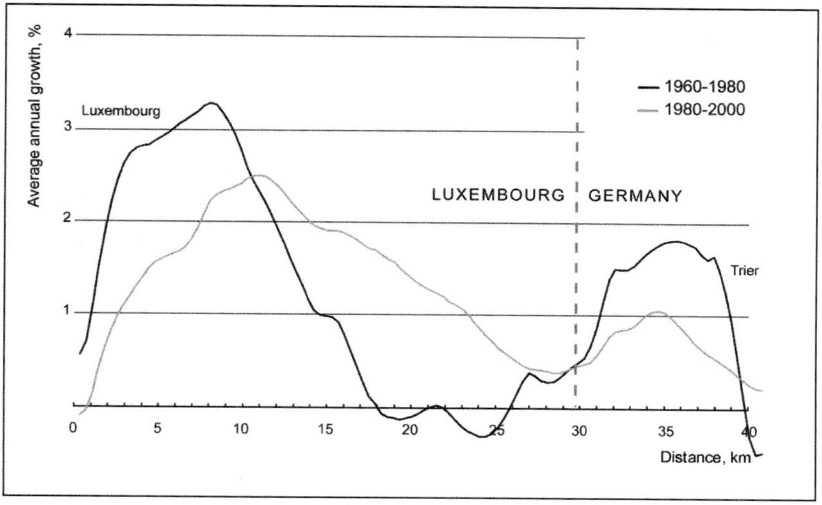

Calculations: Walther and Dickes.

From the cartographic point of view, the analysis of the rates of demographic growth conducted using the local indicators of spatial association show two specific characteristics of the metropolitan areas examined: centre-periphery effects and/or linear effects linked to national borders (Map 7.2).

In Luxembourg, these two effects are manifested in particular between 1960 and 2000. On the one hand, the profile (high-low) of Luxembourg City confirms that the demographic growth primarily benefited municipalities within the agglomeration and the first suburban ring of Luxembourg, whose increase was higher than average and which have a similar demographic growth profile (high-high). These findings confirm those of other studies already conducted in Luxembourg, which show the scale of the suburbanisation of the population and the mobility that accompanies this (Lord and Gerber 2009, Carpentier 2007), as well as the process of the decentralisation of employment that is least susceptible to agglomeration effects (Walther and Dautel 2010). On the other hand, for the example of Luxembourg, a very marked effect can also be observed between 1960 and 1980, and between 1980 and 2000, indicating that the demographic differentials between Luxembourg and France have been persistent (Map 7.3).

These differentials consist of a cluster whose demographic evolution has been negatively affected for the past 40 years; these are found in the industrial and mining region running from Longwy to the Moselle valley. This group of French municipalities has a totally different profile to that of the same industrial area in Luxembourg, located around Esch-sur-Alzette, and for which the local Moran's I index is not statistically significant. These insignificant values are linked to the fact that these Luxembourg municipalities initially experienced a relative demographic decline, followed by more recent growth. This differentiates them from both the suburban area of Luxembourg, which has grown continuously at an above-average rate, and the French municipalities, whose growth has always been below that of the average on this part of the border.

The cases of Basel and Geneva also exemplify the phenomenon of suburbanisation. In Basel, the peripheral area in the south of the canton that shows a similar positive development (high-high) are restricted to the Swiss part of the metropolitan area, while the municipalities close to Jura and the north-east of the metropolitan area in Germany have low-low values. In Geneva, demographic integration is more intensive than in Basel, as the group of Swiss and French municipalities in the west of the metropolitan areas (in particular Pays de Gex) have experienced higher than average growth and form a clearly recognisable cluster. The depressed zones are the municipalities furthest from the metropolitan centre, particularly to the south of the metropolitan area and in Haute-Savoie.

Map 7.2 – Moran significance map for population growth, Luxembourg, Basel and Geneva, 1960-2000

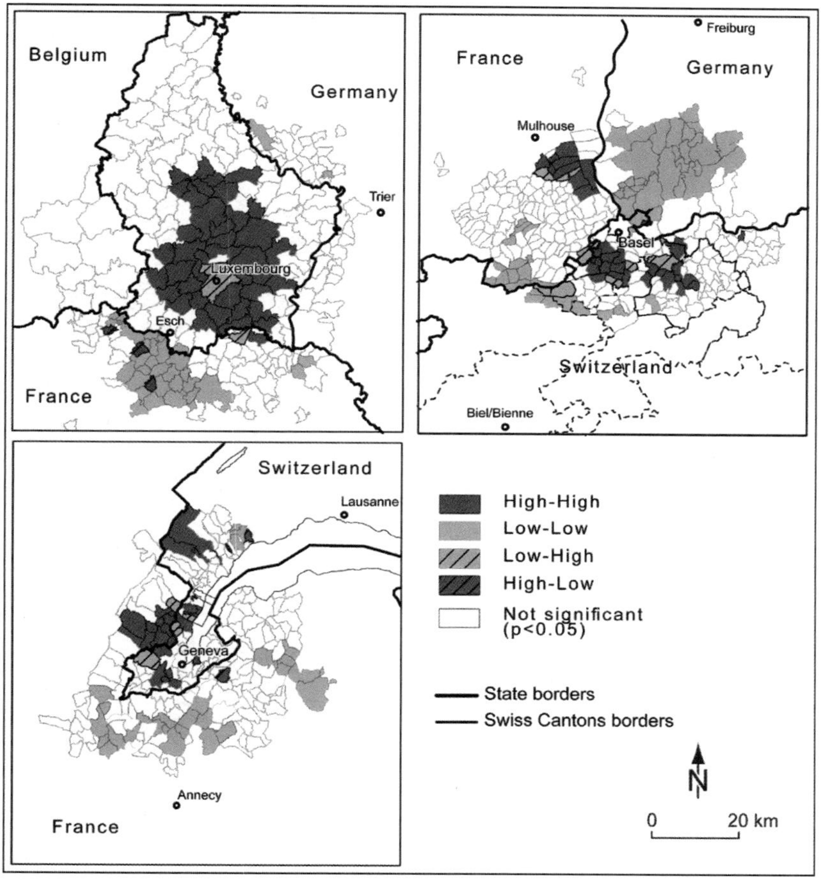

Calculations: Walther and Dickes. Note: contiguity weight matrix is used (queen's case).

The specific temporal characteristics of the process of suburbanisation are especially marked in the case of Luxembourg. Map 7.3 shows that from 1960 to 1980 the group of municipalities in Luxembourg's agglomeration and first ring of suburbs (including the capital and Norstad) experienced above-average rates of demographic growth, in particular contrast to the north of the country, which experienced a rural exodus. During this period, the border division was not especially marked. From 1980 to 2000, peri-urbanisation took hold in Luxembourg's second suburban ring, while Luxembourg City stagnated in demographic terms. A particularly marked division was now seen in the south of the metropolitan area, between the Luxembourg and French municipalities. This division is seen to decrease significantly when one considers the recent period from 2000 to 2008, due to the return of demographic growth to the French municipalities. The Luxembourg metropolitan area is now little marked by clustering, with the exception of the north of Lorraine. This remains an area with very strong growth, which has contributed to shifting the demographic centre of gravity of Lorraine since the end of the 1990s (INSEE 2009a).

Map 7.3 – Moran significance map for population growth, Luxembourg, 1960-1980, 1980-2000, 2000-2008

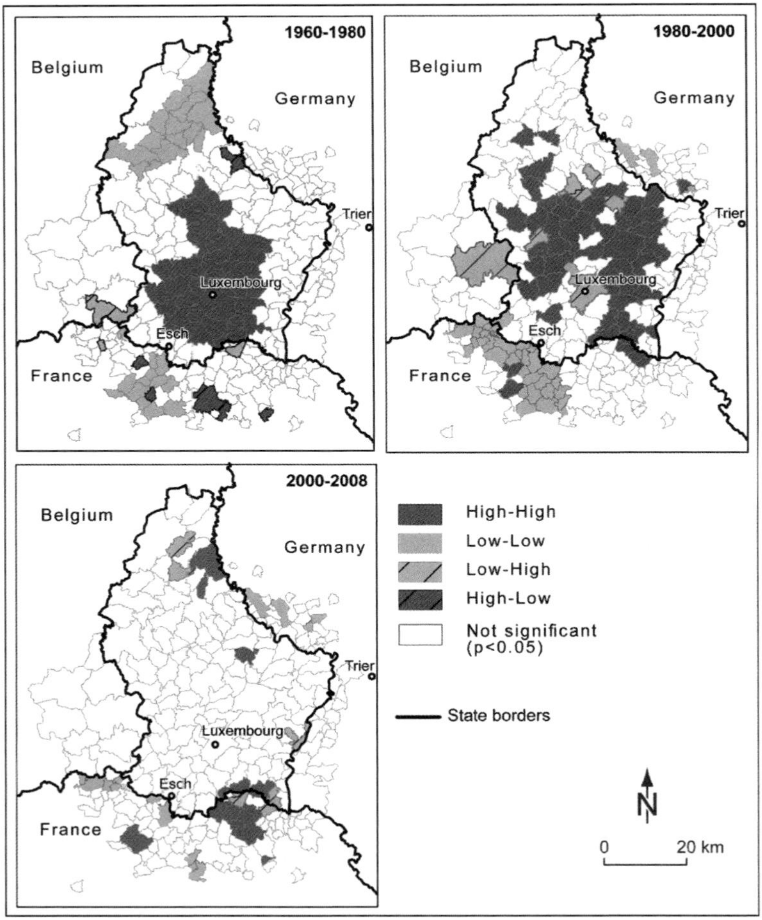

Calculations: Walther and Dickes. Note: contiguity weight matrix is used (queen's case).

Conclusion

Cross-border metropolitan areas share with their European counterparts the characteristic of being subject to intensive spatial change. The processes of the loosening of the urban fabric, and the polarisation that mark these metropolitan spaces take, however, specific forms in the presence of national borders. Although the European project has led to a significant opening up of borders and encouraged exchange between countries, certain demographic differentials can be seen within the metropolitan areas considered in this study. Thus, although Luxembourg, Basel and Geneva are all highly integrated cities from the point of view of the labour market, which is seen in the significant growth of cross-border working, this study shows that the demographic growth is far from being equally homogeneous.

This finding is particularly valid in the case of Luxembourg, all of whose suburban and peri-urban areas experienced spectacular growth from the 1960s onwards, without this being equally true of the neighbouring French, Belgian and German municipalities until 2000. The more recent period appears, in contrast, to show a reversal of this trend, in particular for the eastern parts of the French former industrial region, which experienced a demographic regeneration associated with the spectacular growth in cross-border employment. In Basel and Geneva, the transects appear to show that these two metropolitan areas are more integrated from the demographic point of view. However, the analysis of spatial autocorrelation suggests that in reality the French and German suburbs of Basel have experienced development that differs significantly from that of the Swiss municipalities. These findings show that, according to the criteria adopted in this study, only the western part of the Geneva cross-border metropolitan area (Pays de Gex) appears to be clearly integrated in demographic terms.

As with any other metropolitan region, such a discontinuous demographic pattern can be explained by the extensive use of individual cars, which has allowed work, residency, leisure and education to be detached spatially. In the case of cross-border metropolitan areas, the presence of a national border also offers certain residential advantages, which amplify the present trend towards the concentration of intra-metropolitan spaces. It is thus possible to live in the country that offers the best comparative price and availability of real estate, housing subsidies, or urban amenities, and to continue to hold a job in the country that offers the best salary and career opportunities.

One of the limits of this analysis of cross-border demographic developments is that it only considers density and annual rates of growth as indicators of integration, without taking into account the level of education of the population in question. Recent work conducted on Luxembourg has shown in this regard that all social groups show a trend towards residential decentralisation, but that the highly qualified groups are more strongly represented in Luxembourg City (Lord and Gerber 2009). Among these highly qualified individuals, foreigners have a segregation index and absolute concentration index considerably higher than those for other categories. At the same time, these studies indicate that unskilled Luxembourgers – and to a lesser extent unskilled Europeans – live in the municipalities with the lowest levels of social diversity. From 1991 to 2001, these trends did not particularly intensify. Studies conducted at the level of the functional metropolitan area would allow it to be established whether this stability is a national exception accentuated by the exceptionally high price of real estate in Luxembourg in comparison to the rest of Europe, and would also allow comparison with other European cross-border metropolitan areas.

Note

The author wishes to thank Pascal Dickes for his help in conducting the statistical analysis and the cartography during his internship at CEPS/INSTEAD in August 2008.

PART 3

GOVERNANCE STAKES AND STRATEGIES

CHAPTER 8

Governance of the Greater Region and Cross-border Metropolitan Management

Christian LAMOUR and Franz CLÉMENT

Territorial governance is increasingly the result of negotiations between public and private actors who do not necessarily belong to the same spatial context but who share a common interest at a certain level (Allen and Cochrane 2007). This process means a greater recognition of the non-political sphere in the management of the public sphere. Parallel to the development of this partnership, major debates have been conducted on the best level of intervention for development policies since the 1980s. The region and the city are alternately seen in Europe as the most relevant levels for increasing economic attractiveness and generating social and environmental harmony. Today, the metropolitan region is seen in state and EU policy-making as the key geographical context in terms of stability, as the global economy is based on a persistent network of urban areas concentrating the knowledge economy, orientating informational flows and bringing together command functions (Castells 1989, Scott 2001). The organisation of partnerships centred on competitive metropolitan regions can thus be seen as an effective way of promoting sustainable development.

The European integration project has furthermore given birth to a dual territorial process with, on the one hand, the reinforcement of cross-border functional regions and, on the other hand, the construction of public partnerships within the cross-border areas such as Euroregions in order to facilitate spatial cohesion. The cooperation areas and the functional spaces are not always equivalent. However, this is not a problem in itself as long as transfrontier partnerships can undertake decisive policies at the scale of functional areas. Efficient cross-border governance is not based on the unification of functional areas and institutional jurisdictions, but rather on the ability of a certain number of decision-makers to define a project at a given spatial level when there is a common interest across borders.

From the start of the 1980s, Luxembourg and the surrounding regions have created a Euroregion known as the "Greater Region" (Map 8.1). With its 11 million inhabitants and 65,000 km², this cross-border territorial grouping does not conform to the spatial limits of the cross-border metropolisation driven by the Luxembourg capital (see Decoville and Sohn, Chapter 4). However, it regroups the public authorities that are able to take action with regard to the attractiveness of the city-region. It is thus important to investigate the potential role played by this Euroregional network in the organisation of this cross-border space. Is there a Greater Region approach to metropolitan issues? If so, what is the role played by this cross-border territorial grouping?

Map 8.1 – Territory of the Greater Region and urban polarity

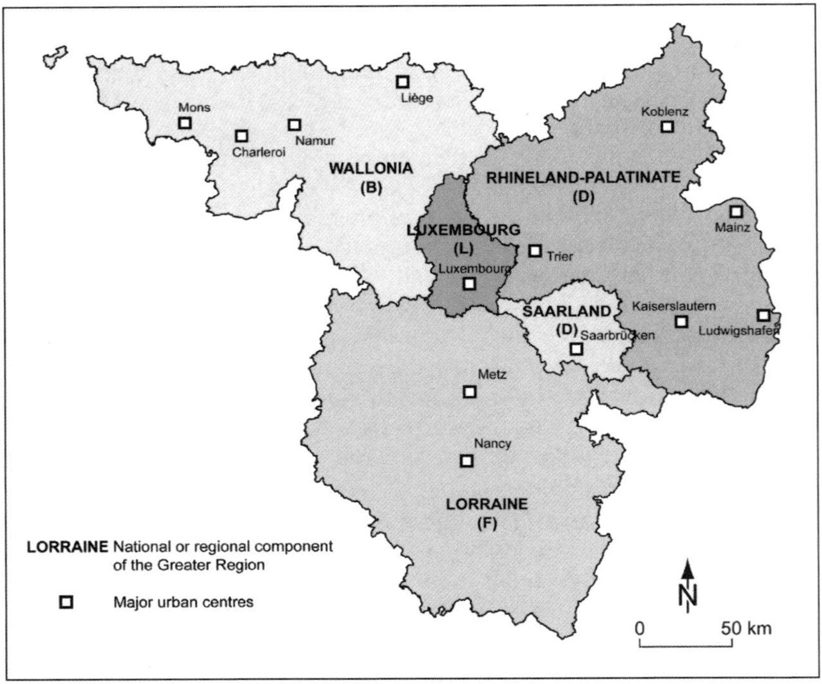

Author: Lamour, 2010.

The central hypothesis is that the Greater Region contributes to reinforcing a multi-level web of actors involved in transfrontier metropolitan development. The national and regional authorities brought together within the Euroregion initiate multi-polar governance via strategic discourse, shared expertise and specific funds supporting the interaction of urban actors.

The first part of this chapter examines the development of Euro-regional and cross-border metropolitan governances and the connection between the two processes within EU policy-making. It is followed by a presentation of the institutions and areas of cooperation established within the Greater Region in order to show the issues at stake in the partnership, but also the territorial complexity of the collaborations. Finally, an analysis of the strategies and reports concerning the work initiated by the executives of the Euroregion since 1995 will provide clarification of the role played by this political system in promoting cross-border metropolitan governance.

Euroregional and metropolitan governances: the networks of actors embedded in the history of development strategies

The cross-border territorial projects result from a Europeanisation of development policy and an expanded vision of spatial organisation. This first section emphasises the territorial logics at work, the academic critiques concerning the efficiency of established governance networks, and also the role of these new assemblages as receptacles of European norms and values concerning metropolisation and polycentrism.

Euroregions – criticised but perpetuated

The European authorities have not been delegated powers by the states in terms of territorial management. However, they play a central role in the orientation of European policies of spatial cohesion. The development of Euroregions is thus strongly correlated to the first strategies promoted by the EU. The cross-border regional institutions blossomed at that time, as they were the prototypes of the "Europe of the Regions" and were also part of the logic of a "Europe without Borders".

In 2003 there were over 70 cross-border regions, whose size and level of cooperation vary a great deal. As shown by Perkmann (2003), these territorial groupings are both driven by public actors and linked to EU regional policies. Far from imperilling the sovereignty of nations, these initiatives contribute, above all, to generating multi-level territorial governance between the European institutions, the states and the local authorities. However, the cross-border regionalism analysed by several researchers often appears as a bureaucratic construction from which concrete results do not always materialise (Scott 1999, van der Velde and van Houtum 2000). Criticisms have been made concerning the capacity of these projects to create popular engagement (Strüver 2004). Furthermore, in the absence of a transfer of powers from national authorities to cross-border public bodies, the lack of power can be seen as the key problem of the Euroregions (Blatter 1996, Hansen and Serin 2007).

Despite these criticisms, the Euroregions have not disappeared. One of the reasons advanced for their durability relates to their importance from an informational and relational point of view. The cross-border institutions are powerful tools, not in the development of concrete structural policies, but in the production of knowledge, the definition of common interests beyond the limits of states and the slow emergence of power understood as the collective capacity to act on a space within several national institutional structures. The actors in question want to remain within these institutional systems without strong powers because they facilitate informational transfers between states at low cost, contributing to the strategic positioning of each associated actor within their immediate European environment. Also, the Euroregions are systems that potentially intensify the flows of information between the institutions located on the same side of the border (Grix and Knowles 2003). Long-term Euroregional cooperation can contribute to communication and improved relations within a state, as it involves national coordination to deal calmly with cross-border exchanges. Finally, the Euroregional field of debate can constitute a platform for the spread of the EU's values and objectives (Scott 2002). In particular, this can be the case in relation to the issue of metropolitanism and polycentralism.

Cross-border metropolisation and the polycentric temptation

Since the end of the 1990s the European institutions have promoted the idea of polycentric urban regions to improve the connections between cities and their peripheral regions (Scott 2002). Polycentrism appears as a means to overcome the centre-periphery problem within each region. It allows a mode of regional development based on a network of cities. Successive EU strategic documents have supported this idea of regional development. The European Spatial Development Perspective (European Commission 1999), followed by the Leipzig Charter on Sustainable European Cities (European Union 2007a) and the Territorial Agenda of the EU (European Union 2007b) emphasise polycentric urban cooperation. The Territorial Agenda explicitly states that it is necessary to establish cooperative projects within cross-border urban regions.

Polycentrism has two complementary aspects. The first is morphological and relates to the distribution of urban centres over a given territory (hierarchy, positioning, number of cities). The second is relational and informs the multidimensional flows between cities. A multipolar situation involves the existence of functional complementarities. It requires both the presence of structural interdependencies, in particular in economic terms, but also in terms of collaboration (Glanzmann *et al.* 2004). In the absence of economic interaction and functional complementarity between cities located close to each other, it is important to question the ability of networks of actors to structure polycentric urban regions (Davoudi 2003). The will to carry out multipolar development on its own does not appear to be sufficient for the latter to establish itself.

The concept of polycentrism, used initially to understand urban organisation, has been reused to establish normative European strategies for regional development transferred and debated in the field of Euroregional governance. Polycentric metropolitan development has thus been added to the list of "border issues" that actors engaged in the fields of transnational debate can use as a point of reference in their interactions (Häkli 2009). It is of potential interest to political decision-makers managing medium-sized urban centres relatively close to one another and eager to conceive a joint attractiveness policy.

Before addressing the manner in which the Greater Region has been able to adopt the issue of polycentrism, it is important to have a historical and geographical understanding of this Euroregion.

The Greater Region: the European Union on a $1/64^{th}$ scale?

Far from being a simple collection of institutional territories, the Greater Region is a political arena in which a series of key actors address spatial issues at different scales within the Euroregion. To demonstrate this shifting reality, the development of the institutional structure and the perimeters of cooperation must be examined. Moreover, the opening up of the cross-border debate to civil society and its consequences for governance within the Greater Region, are also addressed. Finally, taking into account these changes allows the issue of the limits of governance of the Greater Region to be raised.

Increasingly fragmented executive power and territorial differentiation

The legal act that founded the cooperation within the Greater Region is undoubtedly the French-German-Luxembourgish cooperation agreement signed in Bonn on October 16, 1980. At that time, cross-border cooperation in Europe was essentially a matter for nation states; it is thus not surprising that the co-signatories were the governments of France, Luxembourg and Germany. The cooperation was instituted with the aim of consolidating and developing relations between neighbours in matters of common interest, notably in the administrative, technical, social, economic and cultural domains. Two institutions were established to ensure joint work. First, there is a structure known as the "intergovernmental commission" whose role is to develop recommendations and, where appropriate, specific joint projects. It is supported by a "regional commission" that submits cross-border recommendations and projects. The members of these two commissions are different. The Luxembourg central government is the only one represented in these organisations. On the French side, the state is represented by various prefects (the Prefect of the Moselle and the Lorraine region and his counterparts in charge of the departments of Meurthe-et-Moselle, Meuse and Vosges). The situation is more complex on the German side. The Federal government sits on the intergovernmental commission, while the German states of Saarland and Rhineland-Palatinate are found in the regional structure. Furthermore, the zone of cooperation is not a mere assemblage of national/regional territories of participative jurisdictional actors. In addition to Saarland, only a part of Rhineland-Palatinate is included on the German side. The Lorraine region is the area of cooperation on the French side but it is the departmental level that is officially emphasised. That level of French state authority is represented within the regional commission. This is indicative of an era prior to the decentralisation and regionalisation of French public policy, which favoured the division of the national territory into departmental areas (Map 8.2).

Map 8.2 – Institutional powers and the area of cooperation within the Greater Region in 1980 (see colour section)

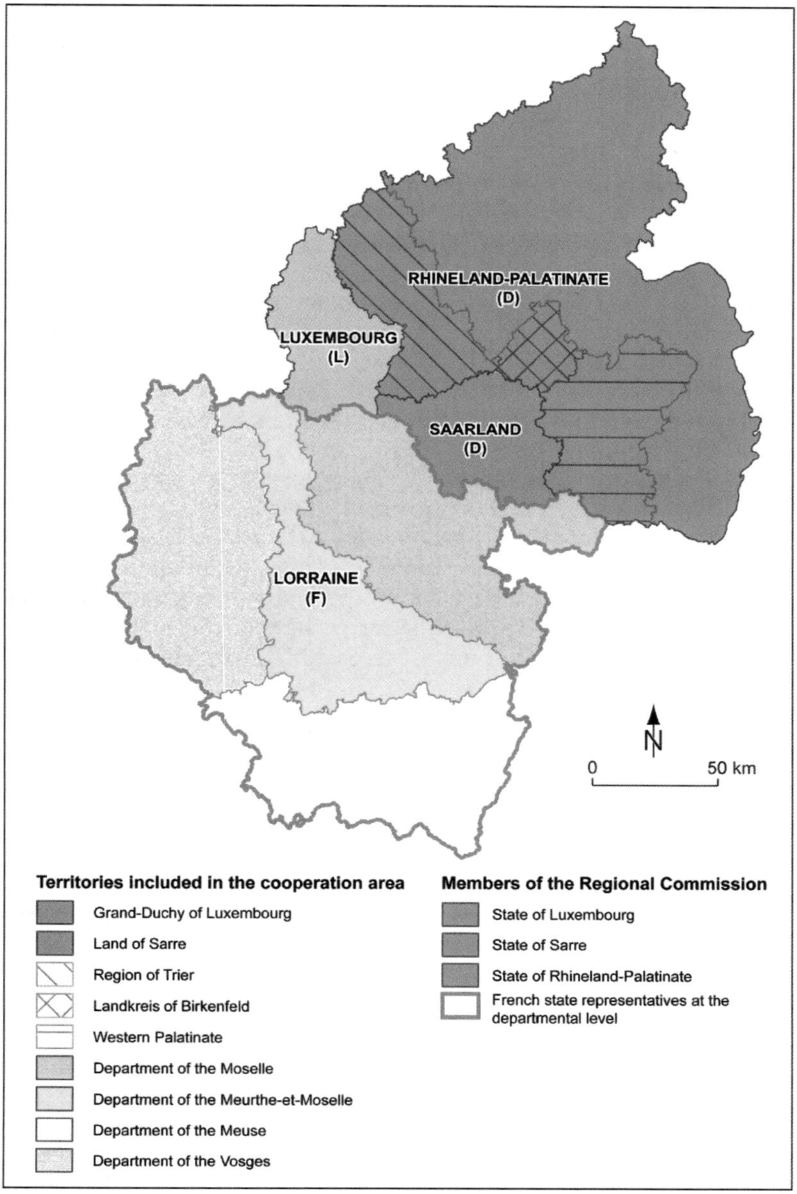

Author: Lamour, 2010.

The 1980s were a turning point in Europe regarding territorial organisation. Several states emphasised the process of power sharing at the European level and also transferred powers to local authorities (Bäck, Heinelt and Magnier 2006). The process of European unification and state decentralisation was accompanied by an in-depth debate in many parliamentary spheres. This had a clear impact on the structure of cross-border collaboration. At the level of the Greater Region, from 1986 onwards, it was decided to create an Interregional Parliamentary Council (Conseil parlementaire interrégional, CPI), allowing the progressive reallocation of power. On the Luxembourg side, the parliamentary representation is national. The German side consists of parliamentarians from the two states represented in the regional commission. The creation of the CPI resulted in changes primarily in terms of the French representation, as the new structure included the Lorraine regional council, which was not included in the two commissions established in 1980. Finally, this new organisation saw the inclusion of Belgium in the collaboration, via three units: Wallonia, the French-speaking Community and, gradually, the German-speaking one. The CPI is a consultative assembly aiming to promote the economic, social and cultural role of the Greater Region, and to contribute to the development of cross-border cooperation. It produces recommendations on the basis of work done internally by six commissions: economic affairs; social affairs; transport and communications; environment and agriculture; education, training, research and culture; and internal security. Its recommendations cannot be issued against the will of any of the member national parliamentary delegations[1].

1 *Convention relative à la création d'un Conseil parlementaire interrégional* [Agreement relating to the creation of a CPI], 1986; *Règlement d'ordre intérieur du Conseil parlementaire interrégional* [By-laws of the CPI].

Map 8.3 – Institutional powers associated with Greater Region cooperation in 1995 (see colour section)

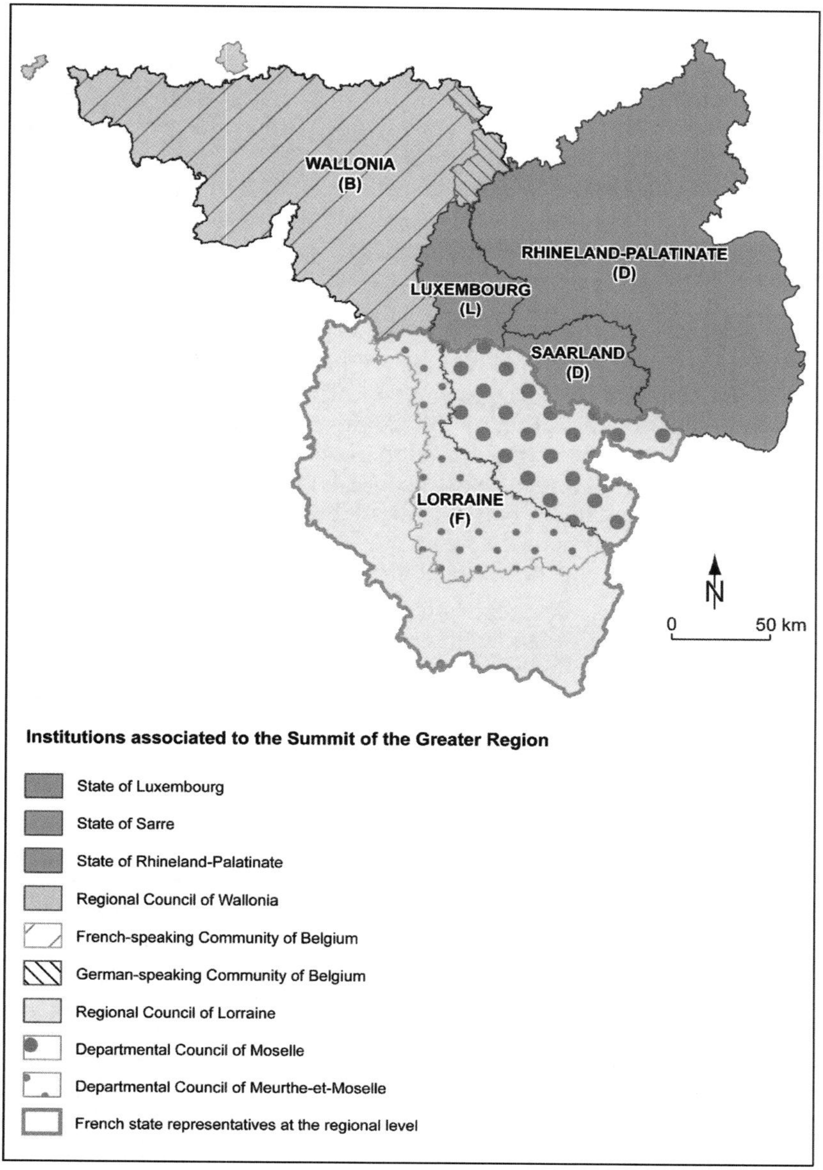

Author: Lamour, 2010.

This development led to the establishment in 1995 of the "Greater Region Summit", a new initiative bringing together the executives of the CPI members and also the French general councils of Moselle and Meurthe-et-Moselle. In contrast to the regional and intergovernmental commissions, the French state (represented exclusively by the Regional Prefect) participates in deliberations but does not vote on resolutions. The German federal government does not participate in the Summit (Map 8.3). For 10 years, this group of political actors charged with promoting cross-border cooperation within the Greater Region existed in parallel with the intergovernmental and regional commissions. In 2005 an organisational fusion took place. Strategic guidance was entrusted to the "Summit of Executives", which included the authorities represented from 1995 onwards. The Summit became the body responsible for pushing forward cooperation. This led to both the spatial enlargement of the institution with the joining of three Belgian administrative entities, and also a fragmentation of the Euroregional executive power. In 1980 the direction of collaboration was defined by just three countries, while in 2005 ten public authorities of different levels (central states, regional authorities, local councils) collectively defined the direction of cross-border policy. It will be noted, however, that the political representatives of the main cities of the Greater Region remain excluded from this dynamic of institutional integration.

In parallel to these developments, the European Interreg programmes were established in order to facilitate cross-border cooperation. This had a considerable impact on the spatial limits of operational cooperation. In 1980 the "cooperation area" was relatively unified, with a single "Saar-Lor-Lux-Trier/Western Palatinate" zone. Interreg divided the cooperation zones, seeking proximity across borders. Three areas eligible for European funding emerged: Wallonia-Lorraine-Luxembourg; Germany-Luxembourg; and Saar-Moselle-Western Palatinate. These areas can overlap and within each grouping, certain parts, referred to as "adjacent", cannot benefit from EU funding at the full level. It was not until the Interreg III 2000-2006 programme that EU funding was available at the level of the whole Greater Region (Map 8.4). The Interreg IVA 2007-2013 programme led to new areas of cooperation. The Greater Region Interreg zoning excluded some part of the Euroregion like the northern part of Rhineland-Palatinate in Germany and the whole of Wallonia at the exception of the Province of Luxembourg and Liège (Map 8.5).

Map 8.4 – Areas of cooperation within the Greater Region, 2000-2006 (see colour section)

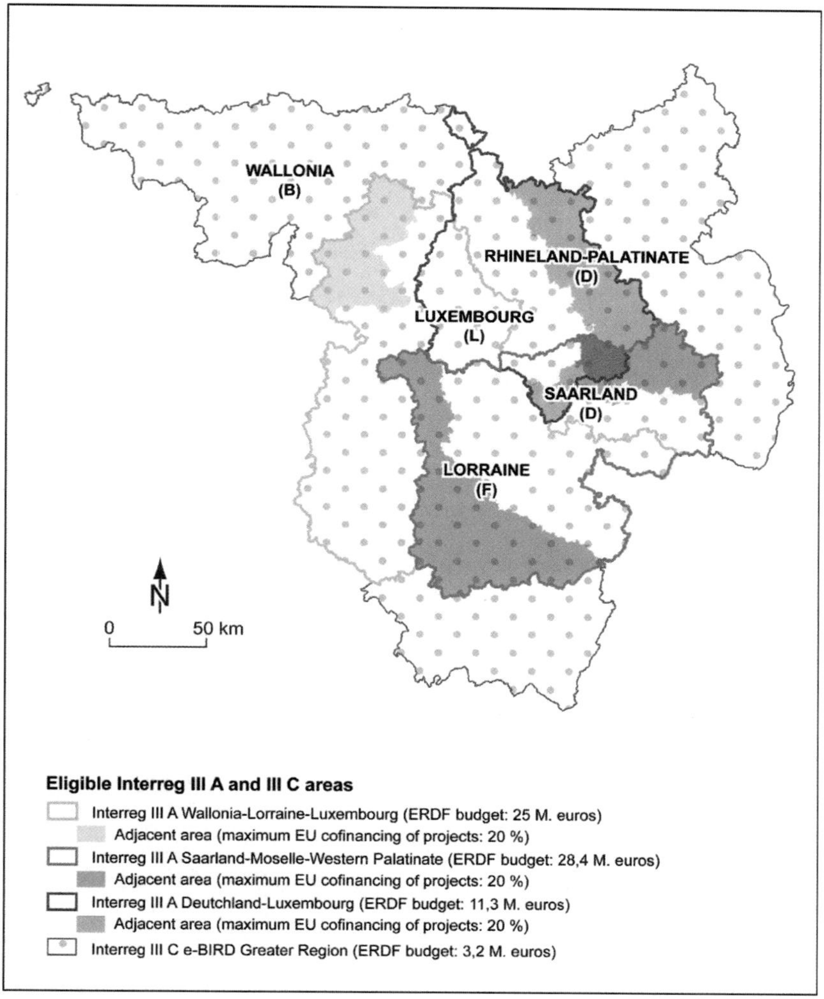

Author: Lamour, 2010.

Map 8.5 – Areas of cooperation within the Greater Region, 2007-2013 (see colour section)

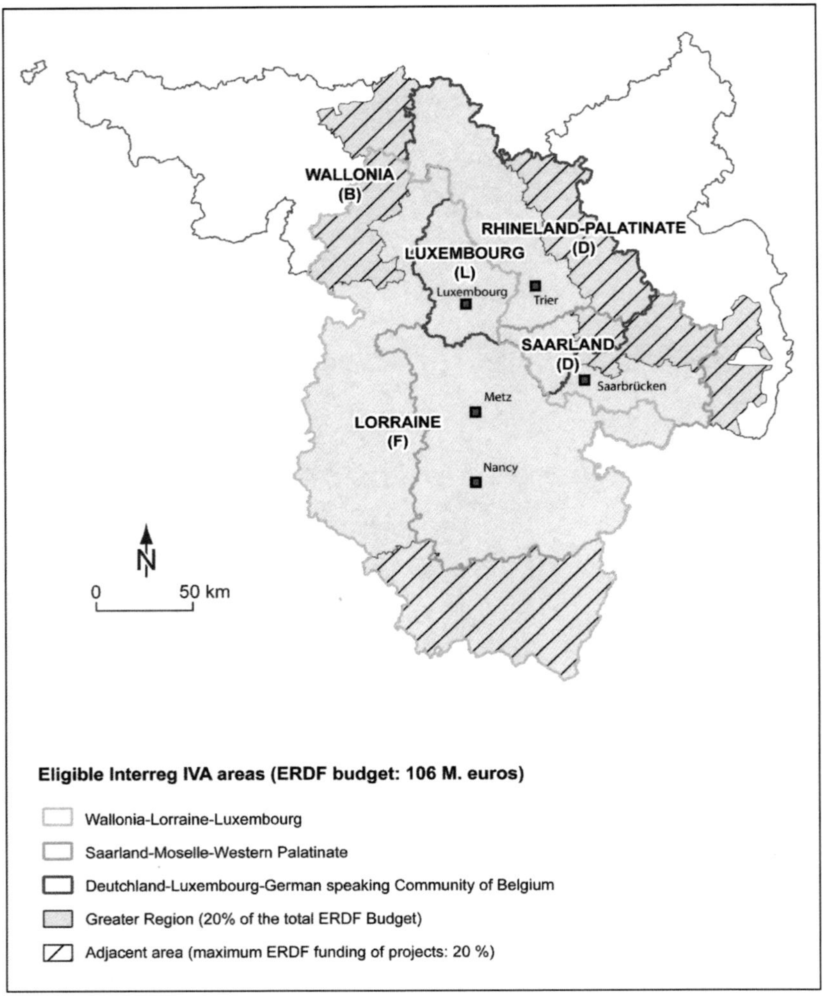

Author: Lamour, 2010.

A formal opening to civil society

In 1997 an Economic and Social Committee for the Greater Region (Conseil économique et social de la Grande Région, CESGR) was established by the Summit. This initiative formalised the inclusion of civil society within the cross-border debate. The CSEGR is a consultative body whose mission is to produce recommendations related to the economic, social, cultural and territorial development of the Euroregion. It also contributes to social dialogue by referring to the recommendations and experiences of its constitutive members. Its work is closely tied to that of the Summit. The committee is assigned tasks by the executives of the Greater Region. It therefore reports to them on its activities and can propose to them issues it wishes to examine. Since its creation, 30 resolutions have been issued for the attention of the Summit by this institution on issues such as the removal of administrative obstacles, transport, clustering, employment and bilingualism. Like the CPI, the Economic and Social Committee for the Greater Region has set up working groups on specific topics such as small and medium-sized businesses, public transport and infrastructure, professional training and health. The civil society representation within this body is divided into three distinct groups, occupying an equal number of posts: workers' organisations, employers' organisations, and other bodies appointed by each member region. There are a total of 36 members distributed equally between the six states/regions represented on the committee: Luxembourg, the Lorraine region, the Walloon region, the German-speaking community of Belgium, the state of Saarland and the state of Rhineland-Palatinate. In contrast to the CPI and the Summit of Executives, the CESGR stipulates in its bylaws that its decisions are taken by a majority of votes cast. It is, however, also specified that everything is to be done to reach as wide a consensus as possible. In addition, minority opinions are taken into account in resolutions when expressed by at least six incumbent members, this being the equivalent of a national/regional delegation.

Concrete examples resulting from the work of the Economic and Social Committee include two bodies concerning civil society. First, there is the Interregional Employment Market Observatory, a body proposed by the CESGR to the Summit, which decided to officially create it on the basis of a network of institutions nominated by each national partner of the Greater Region. Starting in 2001, the observatory has collected and produced data allowing the elected representatives of the Euroregion to better understand the employment market of the cross-border area and to define joint resolutions. The creation of this network, composed of six specialised institutions, serves as an integration of civil society, legitimated by its expertise. The second example relates to public mediation. In 2003, following a proposal made by the CESGR, the Summit created a network of public mediators whose role consists of facilitating the daily life of citizens who must face the legislative fragmentation of the Euroregion. There is thus a clear desire to provide transparency and bring supervisory bodies closer to citizens.

A zone of cooperation and not of territorial integration

The evolving institutional structure of the Greater Region reflects both the deepening of European intergovernmental governance and the increasing opening up of decision-making networks toward civil society. However, the Euroregional initiatives that bring together national executives are not accompanied by any transfer of powers between national and Euroregional authorities. Also, the CPI is not directly elected by citizens and cannot issue resolutions contrary to the opinion of one of its national representatives. Furthermore, it never has anything other than a consultative role. It is thus not comparable to the European or national parliaments (Clément 2008). The Summit of Executives, like the Regional Commission before it, is above all a bringing together of a political community, using a collection of networks of interconnected actors to reinforce the membership of a common area. What are involved are flexible links generating information, debate, participation and specific commitments at multiple levels. Furthermore, the Summit and the CPI are assemblies that seldom meet, in comparison to national executive and parliamentary structures (every 18 months for the former and at least once a year for the latter). However, this very spread-out timetable of meetings is ultimately only the formal seal on work done and the outlines of work to come. It is important to note that the dialogue is held via specific programmes and actions between Summits.

At the end of the day, the Greater Region epitomises the emergence, problems and opportunities of an interregional mode of governance that relies on flexible policy networks in order to resolve the problems of territorial cohesion. Although this system of partnership is not a simple one within a national framework, complexity increases when the area to be managed includes multiple states. These difficulties do not, however, mean that structures of cross-border cooperation, in particular the Greater Region, are unable to influence the structuring of networks of actors and the promotion of new territorial visions.

The emergence of a polycentric metropolitan area via multiple interconnected networks of cooperation

In this final section, an examination of the discourse and actions taken or initiated by the executives of the Greater Region aims to analyse the role of this intergovernmental structure in the emergence of cross-border metropolitan governance. To this end, the political declarations made at the Summits, the networks of expertise put in place and the impact of the European projects are each analysed in turn.

Joint declarations in favour of a multipolar urban organisation

The joint declarations of the Greater Region Summit support a metropolitan and multipolar organisation. First, they put a lot of emphasis on highways, logistic nodes and fast public transport connections between urban poles. From the 1st Summit in 1996, the connection of Luxembourg to the TGV high-speed rail network and an improved cross-border road network have been central to the joint recommendations. Progressively, multimodal freight (logistics platforms) and passenger transportation have been debated. A rail link between Trier and Luxembourg-City was brought up at the 3rd Summit in 1997. The project of a rapid rail connection between Brussels, Luxembourg and Strasbourg was also given priority. There are thus plans to improve the accessibility of Luxembourg's capital city and to better connect it to the surrounding urban poles.

Territorial management is also an issue constantly addressed at the Summits. The Euroregion's political actors share in this field the strategic approach defined in Brussels in the 1990s: the fundamental idea that urban regions rather than individual cities are a good level at which to define territorial competitiveness. At the 1st Summit of the Greater Region, in 1995, the participants recommended "the definition of a development strategy for urban networks… in accordance with the work of the European Union"[2]. The following year this position was reaffirmed more strongly at the 2nd Summit. A joint spatial development strategy was called for which "would deepen reflection on the creation of a more balanced European urban network"[3]. The declaration of the 3rd Summit went even further in terms of its spatial visions. The participants affirmed that there were points of agreement between them concerning the "creation of an integrated transnational metropolitan system"[4]. The focus on urban organisation reached its height at the 10th Summit in 2008. It was announced that the next meeting of executives would concentrate in particular on territorial planning, relying in the short term on an urban network of major cities. In the longer term, the creation of a cross-border polycentric metropolitan region capable of competing with the major national metropolitan regions was also called for (see Sohn and Walther, Chapter 11). The building of such a strategic vision is based on an applied research project entitled Metroborder and

2 Déclaration Commune Finale du Sommet de la Grande Région [Final joint declaration of the Summit of the Greater Region], 1995, p. 4.

3 Déclaration Commune Finale du Sommet de la Grande Région [Final joint declaration of the Summit of the Greater Region], 1996, p. 6.

4 Déclaration Commune Finale du Sommet de la Grande Région [Final joint declaration of the Summit of the Greater Region], 1997, p. 7.

funded by the European Spatial Planning Observation Network (ESPON). In addition to the creation of a multi-level partnership between the national and regional authorities of the Greater Region, two committees including the cities linked within the Quattropole and Lela+ urban networks have been put in place. These Metroborder committees, coordinated by the Luxembourg state, constitute the only body bringing together all institutional territorial powers whose concerted action may be decisive in terms of transfrontier metropolitan attractiveness. The Greater Region and, through it, the Luxembourg state, thus occupy a leadership position in relation to the initiation of multi-level governance focusing on the cross-border polycentric urban region. There is no calling into question of city networks, which will continue with their programmes; rather, connections will be made between decision-making spheres at different territorial levels.

Expertise reinforcing metropolitan structuring perspectives

Knowledge, more than funding, constitutes a key policy-making resource (Saurugger 2003). The Greater Region, via the Summits, aimed to develop expertise to enrich public debate and set the broad outlines of a joint policy. This knowledge allowed joint understanding via a corpus of terms, concepts and findings. The progressive creation of Euro-regional knowledge is even more important, as informational flows are developing between actors not sharing the same language or culture and thus not intuitively sharing the same analytical and planning approaches.

This expertise has taken two forms: research contracts and the creation of formal analysis networks. The contracts are generally designed to provide results for the preparation for Summits. During the preparation for the 2nd Summit, a study was established to define the factors hindering cross-border mobility in matters of employment and training and the means to facilitate professional exchange. Following this study, the 3rd Summit initiated the creation of the Interregional Employment Market Observatory. The 4th Summit set in motion the progressive creation of a database by the national statistical offices of the Greater Region working together. There is thus a permanent structure of multi-polar expertise developing over the course of the Summits. Employment and economic issues constitute the major areas of interest. The Greater Region has, for instance, carried out a study on economic clustering to better understand the interregional environment and define collective vision on that issue[5].

---

5 Social and Economic Committee of the Greater Region, 2006.

In addition, the 7th Summit of the Greater Region mandated a political commission to design a collective vision for the year 2020[6]. This group has made proposals in eight areas that are regularly at the centre of debate: culture; education/vocational training; research; economics/employment; social networks; transport; environment/territorial development; and institutionalisation. The working group provides a summary of the current situation, projects of limited scope that can be carried out in the short term and the organisations that allow the ten-year targets to be achieved. Research centres, laboratories, the University of the Greater Region, economic institutes, marketing agencies, the European transport and logistics academy are conceived as cross-border polycentric institutions contributing to the target aims. In addition to the promotion of these multipolar thematic agencies, the commission advocates reinforcing the connections between urban centres. It is envisaged that collaboration will be established, which will facilitate the interconnection of actors in charge of high-level functions (health, universities, theatre and sports facilities). This multi-level strategy does not entirely fall within the issues of "territorial/environmental development", but also within the scope of the institutional organisation of the Greater Region. An Interregional Council is seen as an executive body to replace the Summits by 2020. This council will involve the mayors of the main cities. What is implicitly being advocated is thus an enlargement of the Greater Region partnership system. Of course, all this is a plan for the future, but the creation of permanent Metroborder committees, linking the cities and national and/or regional public authorities of the Euroregion, can be seen as a first step to bring together the city councils. Their action will be decisive for the definition of urban polarity and the higher supervisory bodies also aiming to make the territory attractive through their public policies.

Increasing the density of urban networks via European programmes

In addition to planning a polycentric metropolitan organisation via official strategies and expertise, the Summits of the Greater Region have also developed budgetary programmes to promote the emergence of thematic networks. These funds facilitate transfrontier ties between stakeholders generating new information flows, strategies and specific commitments.

One of the most emblematic cases is the "Luxembourg and Greater Region European Capital of Culture 2007" project. Usually, this cultural event promoted on a European scale involves a single city. Luxembourg,

6 "Vision d'avenir 2020" policy commission, 2003.

however, proposed to its Greater Region partners that they put in place a major regional project at the 5th Summit in 2000, a proposal that was received favourably and ultimately accepted by the European institutions. This initiative allowed the creation of cultural attractions within an urban and polycentric framework. This territorial enlargement certainly increased the visibility of Luxembourg and its capital at an international level, as well as creating cross-border cultural connections. A quarter of the projects established within the framework of the Year of Culture – 139 in total – were cross-border ones (Varbanova 2008). These projects allowed the major cultural and urban centres within the Greater Region to be connected. Luxembourg, Metz, Nancy, Trier and Saarbrücken are often involved in cross-border actions (Sohn 2009a). This programme created opportunities favourable to the emergence of a polycentric metropolitan area in the field of culture. The process was even more interesting in that certain projects connected major centres with smaller towns located on the other side of the border. Certainly, the events were mostly concentrated in Luxembourg and the funds provided by the peripheral regions to finance the projects were far inferior to those provided in Luxembourg, but the operation can be seen as an experiment in cultural coordination between supervisory authorities, city councils and a series of civil society actors via concrete actions and decision-making committees. The results were convincing for the Greater Region partners, which is why the Euroregion Summit created a new association in 2008 entitled the Greater Region Cultural Space, charged with promoting joint actions in this field. This includes the national/regional supervisory authorities, as well as the cities of Luxembourg and Trier.

In parallel to this cultural work, now rendered long-term, the Summit of the Greater Region has obtained, within the framework of the Interreg III C programme, a specific budget for the Euroregion: the e-Bird Regional Framework Operation (RFO). This initiative has enabled new strategic approaches via projects involving actors in the four countries of the Euroregion. A total of 20 cross-border projects have emerged. The operation has generated partnerships in terms of public debate, studies, cross-border visibility and the virtual community (in particular thanks to internet sites). In the same way as the "European Capital of Culture", the networks created have involved stakeholders located in the urban poles of the Greater Region. The programme has been designed along three major axes. The first, entitled "economic, social and territorial development", contains ten projects. It is above all focused on issues of cross-border employment, interregional economic flows and technological transfers from one region to another, three topics connected in part to the polycentric metropolitan issue. The second axis, concerning culture and knowledge, includes four projects. The main aim is to promote virtual cultural polycentrism via the internet. Finally, the last axis, called "education, training and research", promotes relationships between actors in charge of higher education, a pillar of urban attractiveness at the international level. Altogether, these projects have directly mobilised 129 structures. The striking fact about these partnerships is the absence of district authorities at the heart of these projects. One primarily finds research and study centres, supervisory authorities at the ministry level and consulates and cultural associations[7]. However, it is difficult to establish precisely the total number of actors in the Greater Region who have been involved in the projects, as the operations have given rise to surveys, symposia, consultations and mass media information. The absence of major city councils in the direct partnerships does not, however, indicate that they have not participated in the formal and informal debates developed through the various projects.

The same strategy of promoting networks of urban actors was instituted by the 9th Summit of the Greater Region (2006), which decided to send a Euroregional proposal to the European Commission for the next Interreg programme. This was received favourably in Brussels and led to the Interreg IV A Greater Region 2007-2013 programme, in which the previous programme units are respected but which reserves 20% of the total budget for projects on the Euroregional level (Map 8.5). Since the start of the Interreg IV A Greater Region programme, 61 projects have been accepted by the selection committee[8]. A high number of approved projects are dealing with environmental issues at different spatial level (e.g. river contracts, water treatment stations, etc.). In addition, there are many actions relating to rural development such as nature park cooperation, green tourism or forestry economy. There are relatively few connections relating to inter-urban development. These take two main forms. The first one concerns the implementation of a master plan for the Saarland-Moselle Est cross-border agglomeration, often described as a developing metropolitan area (the "blue band" project). The second is linked to the knowledge economy and linked either directly to the cities (the "Trilux" science park project between the cities of Trier and Esch-sur-Alzette) or research centres (for example the "TIGRE" project aiming to intensify relations between research institutes and business). The "Greater Region University" project, whose members are located in the major urban centres, is probably the most emblematic of the metropolitan projects in this area. It is located at the intersection of a bottom-up dynamic linking local universities since the 1980s and a top-down policy of supervisory authorities promoting polycentric coordination of research.

---

7 Interreg IIIC e-BIRD. E-Based Interregional development. Réalisations et perspectives [Creation and Perspectives] 2003-2007, 2008.

8 Interreg IV A Greater Region, 2009.

Conclusion

The regional and urban modes of governance developed within the European Union have resulted from a collective will to rethink spatial management in order to ensure territorial cohesion and attractiveness on the international scale. Moreover, the urban and regional levels are increasingly interwoven. The urban region is establishing itself in the EU and national decision-making circles as the right spatial level for competitiveness and sustainability policies. The area of collective commitments is not fixed and unique but more and more flexible and plural. It is important also to note that the functional organisation of space is just one element of territoriality. It is currently emphasised, due to the context of international competition. However, other factors such as socio-cultural practices and history are just as, if not more, important, especially in cross-border areas (Keating and Loughlin 1997). The transformation of function spaces covering multiple states into territories with unitary regulation requires the developing of a political identity going beyond the nation state. European local governance thus requires communications platforms that stimulate shared knowledge, institutionalised social linkages and a time-space gradation of commitments in terms of spatial management.

The executives of the Greater Region, via the Summits and the inter-ministerial meetings, determine the agenda-setting in terms of spatial planning. Influenced by EU policy-making, they promote a polycentric vision of the cross-border metropolitan region. Thus, the "Metroborder" project coordinated by Luxembourg is part of the Greater Region development process. However, the design of a permanent and efficient spatial structuring in favour of a cross-border polycentric metropolitan region requires a well thought out reshuffle of public funding beyond the state. The Interreg programmes, which participate in the creation of networks of urban actors, will probably not be sufficient to provide a foundation for strong partnerships, given the small size and short-term nature of the budgets available. A cross-border transition in terms of the management of public funds can be envisaged. The inter-ministerial links at the Greater Region level in matters of research, transport, territorial development and culture allow one to foresee a potentially more integrated territorial policy (without, however, excluding the dynamics of competition between actors and the intraregional territories of the Euroregion, depending on the issue at hand).

Today, if one takes the four models of cross-border metropolitan governance developed by the OECD (2006b) – governance by project, governance by issue, fragmented governance and consolidated governance – it appears that the practice developed by the executives of the Greater Region combines different forms. The Euroregion has progressively established itself as a consolidated space of cooperation, bringing together all public executives with a role in metropolitan management, with the exception of municipal powers (model 4). The Euroregional governance thus directs a large part of the inter-city networks via the allocation of Interreg funds. In addition, the major municipality, while not associated with the Euroregional executive, participate in the "Metroborder" project, piloted by Luxembourg under the name of "Greater Region". There are still, however, initiatives that are not driven by the Greater Region coalition, in particular the QuattroPole and Lela+ networks, managed directly by the cities. Furthermore, the control and coordination procedures for the Interreg partnerships implemented by the executives of the Euroregion remain very flexible and the number of networks is growing, given the increase in structural funds for cross-border cooperation. The governance of the Greater Region is thus becoming in certain ways a fragmented "Russian doll" form of governance (model 3). Finally, over the last two decades there has been both a project-based approach (e.g. Luxembourg and the Greater Region, European Capital of Culture – model 1) and also a thematic approach via the implementation of sectoral working groups, of which the latest is the cross-border workers taskforce (model 2). The Greater Region is a multi-level network of national powers influenced by community policy-making on spatial management and in a position to carry out differentiated polycentric urban interconnection at the cross-border level in order to increase the attractiveness of its cities. However, we should keep in mind that the different networks put in place have designed above all a polycentric flow of information and a limited number of projects increasing the material and immaterial connections of cities. These partnerships have not transformed the functional integration of the city-region which is still strongly mono-centric in terms of economic development. The City of Luxembourg is still indeed the main pole of employment generating directly or indirectly a major part of cross-border movements.

CHAPTER 9

The Role of City Networks in Cross-border Metropolitan Governance

Eric AUBURTIN

The cross-border cooperation that has been developing for over 30 years within the Greater Region gives priority to new forms of alliances that are not limited to relationships between states but rather increasingly involve various local and regional partners at the interregional level. The construction of cross-border city networks, such as QuattroPole, which has connected Metz, Luxembourg, Trier and Saarbrücken since 2000, or the more recent LELA+ network which has connected the cities of Metz, Thionville, Longwy, Luxembourg, Arlon and Esch-sur-Alzette since 2006, provide an opportunity for these different cities to escape their status as "border towns". They take on a European dimension, which none of them would be able to achieve on their own. Set up by the elected municipal representatives of the cities in question with varying degrees of informality, this process has now been repeated at other decision-making and operational levels, as illustrated by the Metroborder research project, backed by a meeting of Greater Region executives in March 2009 to promote the emergence of a polycentric cross-border metropolitan region.

Given the polysemic nature of the concept of "network" and the multiplicity of forms that city networks can take, several clarifications are needed. For geographers, a network is a set of links or relations with a greater or lesser degree of complexity, shaped by nodes and segments constituting an ensemble of multiple relations that are hierarchical and centralised to varying degrees (Brunet, Ferras and Théry 1993). As topological spaces combining places according to a logic of connectivity, geographical networks (urban networks, transport networks) are based on territories that have served as the support for their establishment (Lévy and Lussault 2003). The network, by connecting places with pathways, creates territory. The connected links transform themselves into a zone or spatial system (Painter 2009). Thus, to investigate metropolitan synergies it is necessary to take into account not only the relations established by the nodes of the network but also the role that may be played by the intermediate, or rather interstitial, spaces involved in the research into the network's potentials and, above all, coherence. Pumain and Offner (1996: 201-203) note that while the urban system is characterised primarily by hierarchical relations between cities within the same politico-administrative territory, a city network instead embodies "forms of voluntary or partial structuration", which do not necessarily depend on cities belonging to the same national or regional territory. These networks can thus bring together cities specialising in a given sector of activity (the financial sector, for example), select cities by size (networks of major cities) or be alliances of cities combining urban centres that are close to each other and complement one another. In this case, it is the similarity of the level or nature of their functions that make interactions between the cities more likely.

In the case of the Greater Region, the networks combine cities in adjacent national territories within the same cross-border area. These city alliances take the specific form of a "club of cities" (Pumain and Offner 1996) which, via the regular meetings of the mayors in question, constitute above all a network of interpersonal exchanges, intended to improve the brand image of each city. However, whatever their form, the networks are intended to produce "added benefit in terms of development", and thus while the hierarchy may be disturbed initially, the aim is to pass a threshold of visibility on the European map of cities, and also to pass a threshold of credibility for economic actors. The development of economies of scale via the complementary specialisation of network centres thus continues to be relevant.

One of the aims of this chapter, by examining the articulation of scales, is to highlight the role of the actors both within the city networks and vis-à-vis the territories with which they interact. This multiscale analysis is based on the observation that the city networks involve different orders of magnitude, which intersect with each other to greater or lesser degrees (Lacoste 2003), whether the European urban network, the cross-border network, the national network or the network of municipalities that constitute the agglomeration, or that of the urban area. The geographical approach is combined with a geopolitical one, which seeks to understand how the relations and logics of power function between actors and territories. At the same time, the question posed is this: to what extent are these cross-border city networks capable of moving beyond the stage of political display resulting from the competition that can exist between them to achieve real metropolitan synergies? Can it be envisaged that these structures will succeed in overcoming the obstacles presented by national borders in relation to legal and organisational matters?

The role of polycentrism in the debate about European territorial management leads us first of all to show the ambiguities inherent in the transfer to the political sphere of a concept that relates primarily to the organisation of a space around the urban centres that constitute it. We will then examine the extent to which certain actors in the Greater Region have reflected this in their territorial planning strategies, before analysing the more specific place of the QuattroPole and LELA+ city networks within the Greater Region, and their ability to offer a new form of territorial governance at the cross-border level.

Polycentrism – a polysemic concept in the service of the multiscalar reorganisation of the territory of Europe

The reunification of Europe after 1989, with the enlargement of the Common Market, drew attention to the necessity of achieving greater balance in the relations between the former European "centre" of the megalopolis and the new European peripheries in the East. Polycentrism is thus invoked in European and national development plans to promote more harmonious and balanced development of the European area, while reinforcing its global competitiveness. However, some authors have drawn attention to both lacunae and contradictions within this approach.

Polycentric development, the new form of European territorial development

Since 1989, the opening of the European Union to the countries of Central and Eastern Europe and the arrival of the Single Market in 1993 have contributed, in terms of spatial management and territorial development, to the transcendence of national limits and thus to the opening of new horizons (O'Dowd 2003). The end of the Cold War and the spread of new information technologies have, furthermore, contributed to an acceleration of the process of globalisation, putting into question the traditional equilibria of the European area. A polycentric organisation based on the emergence of centres of economic integration is developing. This configuration appears designed to favour both the articulation of flows within the EU and openness to the exterior (Pastorel 2007).

The need for an overall vision for the whole of the EU territory has thus gradually become clear. Developed from 1993 onwards by the 15 Member States of the Commission, the European Spatial Development Perspective (ESDP) was approved in 1999 at an informal meeting of the ministers with responsibility for regional planning in Potsdam, making the EU territory an explicit concern of the European project. The adoption of the Territorial Agenda in 2007 made the principle of territorial cohesion an objective of the EU and is intended to also ensure the competitiveness of the European territory at the global level, which will allow its internal rebalancing. European polycentrism is based on city networks as the means of connecting the major zones of the economic integration of tomorrow (Allain, Baudelle and Guy 2003). As noted by Husson (2002), territorial cohesion thus also constitutes an opportunity both to put into practice the principle of subsidiarity, as defined in the European treaties since Maastricht (1992), and to promote a new model for the development of the EU.

Polycentrism, by promoting the networking of actors at the same territorial level, also corresponds to the aim of increasing EU competitiveness, promoting innovation by means of cooperation between metropolitan areas and cities. In the era of the globalisation of flows of goods and services, the challenges of competitiveness and attractiveness can only be met through the promotion of a process of metropolisation based at the global and continental level on the networking of metropolitan centres. This is because they are nodes of communications and because they constitute local productive systems adapted to the needs of the global economy (Veltz 1996).

A contradictory vision of territorial development

However, the idea of polycentrism, as formulated by the ESDP, aims to promote objectives that are still too often thought of in terms of reciprocal exclusivity – territorial cohesion and competiveness – the importance of which was reaffirmed by the Lisbon Strategy. Furthermore, certain actors include among the objectives of polycentrism the social equity and environmental sustainability called for by the Gothenburg Strategy. Davoudi (2003) thus denounces conceptual and normative weakness in order to permit voluntary cooperation between cities and the construction of effective polycentric development. The very definition of the concept appears to be merely approximate, as a polycentric urban region is an urban area containing at least two autonomous municipalities that have functional interconnections between them in a space characterised by reasonable proximity (European Commission 1999). The concept can thus be applied at very different levels of spatial analysis, from the level of the agglomeration to that of an urban region or an entire continent, such as the EU. Moreover, even at the scale of the agglomeration, polycentrism can refer to either spatial development planned in a polynuclear way, or to the result of urban sprawl controlled to a greater or lesser extent.

The functional interdependencies are often difficult to define and measure: often, the only criteria are those concerning economic exchanges in relation to the development of the single market. These interfunctionalities have not, to date, been conclusively observed except for the conurbations. Interdependence is much less marked in other cases. The interactions often appear to be much stronger for cities that are far apart but characterised by a high degree of functional specialisation (for example the network of financial locations), thus benefiting above all those cities already integrated into the globalised economy. In addition, the fact that the concept combines, on the normative level, objectives as diverse as those of competiveness and territorial cohesion also weakens its operational consequences (Catan 2007), especially as European studies like the European Spatial Planning Observation Network (ESPON) do not really specify how this approach can be put into practice (at what level of authority and with which division of responsibilities).

The fact that polycentrism is becoming established as a normative model of territorial planning does not in itself constitute a dead end, so long as it is considered not as the ultimate end, as there are other models of organisation and spatial planning that may prove to be equally efficient. As indicated by Dallenne (2009: 117), three major objections can be formulated: "There is no evidence that the sustainability of development is best assured where dispersion prevails: monocentrism certainly has its drawbacks, but is also a source of economies of scale. Spatial structures are the products of history and cannot be changed by decree. In addition, while France is monocentrically organised at the national level, polycentrism prevails at the regional level, especially in north-eastern France."

It is, however, this spatial vision of territorial organisation that is intended to be applied at the cross-border level to allow border cities, handicapped by the presence of national borders, to have an influence at the European level.

The Greater Region: a form of regional and decentralised polycentric organisation

The Saarland/Lorraine/Luxembourg/Rhineland-Palatinate/Wallonia Greater Region was launched at the first summit of regional executives held in Mondorf-les-Bains on September 20, 1995: it is a new and original cross-border cooperation organisation within an area characterised by an explosion in the flows of cross-border commuters into Luxembourg since the arrival of the Single Market in 1993. It remains an unidentified geopolitical object that brings together, in a manner that is informal to a greater or lesser degree, political officials at different levels of state, regional and federal power and responsibility. Lorraine is thus represented by: the prefect of the region who, in the role of observer, symbolises state authority; the presidents of the Moselle and

Meurthe-et-Moselle General Councils most involved in cross-border activities; and the president of the Lorraine Regional Council. For the other regions, the minister presidents of the Walloon Region, the French and the German-speaking Communities of Belgium represent Wallonia, while the minister presidents of Saarland and Rhineland-Palatinate represent their respective states and the Luxembourgish prime minister represents his government. This heterogeneous institutional grouping represents a new form of governance of cross-border relations, which is also seen in the new possibilities granted by the devolution laws in France (1982, 1992) and the development of the European Single Market (Auburtin 2002).

By moving beyond the limits of the nation state, the Greater Region deepens and legitimises the functioning of networks

Initiated by the French, German and Luxembourgish central governments at the start of the 1970s, cross-border cooperation represents a new opportunity for local actors, to the extent that they can take advantage of a location at the edge of the country, which seems to have few advantages, to attract European and national investments. New forms of territorial cooperation thus gradually develop in a "horizontal" manner (cross-border agglomerations; city networks). They allow bilateral or multilateral relations to develop on the basis of shared interests determined by the actors in question and their needs (for example, management of the post-mining areas in the Saarbrücken-Moselle-Est agglomeration).

This networked structure allows regional actors to strengthen their legitimacy within an upwards-facing structure in relation to their respective government and the European institutions, and to play a lobbying role to obtain, for example, European subsidies. It also validates the use of concepts such as polycentrism at the inter-regional or inter-urban level, within a downwards-facing structure. Finally, it allows the reinforcement, in a horizontal manner, of the coordination necessary to implement sectoral or development policies between actors who are asymmetrical in terms of institutional or economic weight.

At the inter-regional level, the cross-border network of cooperation allows the identity of the project territory to be affirmed in the eyes of external actors as an identifiable site whose critical mass is reinforced. For internal actors, this represents an opportunity to overcome a certain number of institutional obstacles. In the case of Lorraine, the state must take into account the resolutions passed by the region and the *départements* at the Greater Region summits. Even in the federal states, the resurgence of regional differences and specificities – as in the case of Belgium between region and province (Belgian Luxembourg within

Wallonia), or in the case of German between states of different sizes and economic weight (Saarland and Rhineland-Palatinate) – makes cross-border cooperation into a valuable political lever to overcome certain reluctances. Thus, the state of Saarland has made cross-border cooperation a constitutional matter, while Bernard Caprasse[1], governor of the Belgian province of Luxembourg, noted in 2000 the need to construct a cross-border Greater Region that would allow him, as a representative to the Walloon government, to relay the interests of the area it represents to reinforce the Arlon-Longwy-Luxembourg axis within the regional development strategy.

... but as yet has not involved cities to any large extent in its work

The establishment of city networks in Lorraine is a response to this reasoning, especially so as the rivalries between the two metropolises of Metz and Nancy go back centuries. In making an alliance with a close urban centre of similar size, the aim is to increase the attractiveness and influence of the partner cities within a European and globalised context where competition between cities is increased (Le Galès 2003). Thus, although the ESDP plays no programmatic role, as the regional development is not an EU competence, its main conclusions have been taken up in the Saar-Lor-Lux+ development policy (IfLS 2001) that constitutes its regional application, and reference is made to it in the majority of regional or national development policies (Map 9.1). In the European spatial context of the major metropolises, the area of the Greater Region is a hollow space with the lowest densities. The Europe2000+ European transport strategy even treats it as a space for the bypassing or relief of the major European flows, or a recreation region, due to the presence of forests and national parks. This view is only a partial one, as although the Moselle valley does not contain any European-scale metropolises, it is undoubtedly one of the busiest transport corridors at the European level for both goods and passenger traffic (over 80,000 vehicles per day). The A31 motorway, which links Luxembourg, Metz and Nancy, is also not just a European transport axis but also a motorway of local and regional importance, meaning this route is close to saturation point in the section that connects Thionville to the Luxembourgish border, and, to a lesser extent, between Pont-à-Mousson and Nancy.

1 Speech by Bernard Caprasse, Governor of the Belgian province of Luxembourg, Mercuriale 2000.

Map 9.1 – The Greater Region in its European context, 2003 (see colour section)

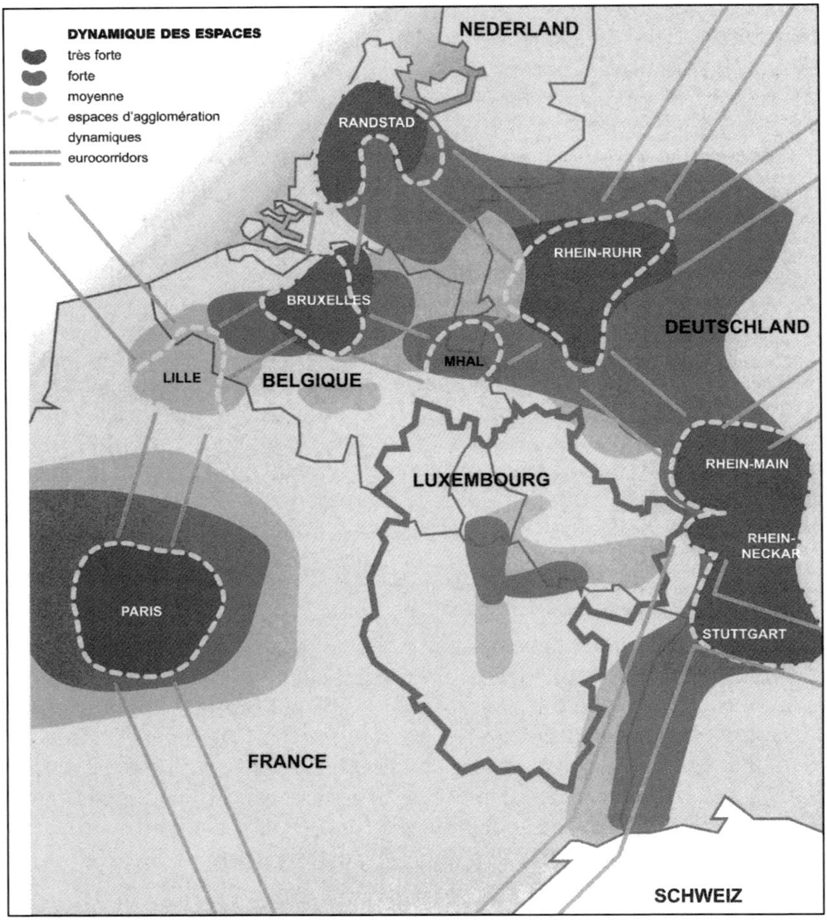

Sources: Ministry of Spatial Planning, Programme directeur d'Aménagement du territoire, Luxembourg, 2003.

The polycentric mode of organisation now appears, within the Greater Region, to be the only one capable of compensating for the lack of a real metropolis of European scale. It thus has a political dimension, as efforts are made to promote a new type of metropolitan governance with inter-urban cooperation as one of the main methods to provide a counterweight to the influence of larger metropolises. Thus, the final Saar-Lor-Lux+ preliminary report (IlSF 2001: 115-118) makes city networks a "priority to promote regional development." The polycentric mode of functioning should allow the position of the cities to be improved vis-à-vis the major axes of communication and centres of research, in order to enjoy improved competitiveness within the European space of the major cities, promoting improved access to city centres via an integrated urban transport and accommodation policy. This could also be achieved by seeking to achieve balanced development of the territory by creating a network of the six largest urban centres (Nancy, Metz, Luxembourg, Trier, Kaiserslautern, Saarbrücken) and the intermediate urban centres such as Arlon, Thionville, Forbach, Sarreguemines and Epinal, for example. This spatial vision of the development of the region via networks thus corresponds to the concept of polycentrism promoted by the ESPON project METROBORDER: the construction of polycentric metropolises within cross-border regions to encourage the rebalancing of the European area. It is particularly in the fields of education and research that metropolitan interactions are most needed to increase the attractiveness of the Greater Region as an area housing high-quality services. The cities must thus take advantage of the complementarities between them to attract businesses and skilled labour. However, the policies of Saarland and Luxembourg are the only ones to make explicit reference to the strategies defined within the framework of the Saar-Lor-Lux+ project, while strengthening the logistical aspect appears to be the most widely-shared objective among the regional and national policies to develop real metropolitan synergy, benefiting in particular from the effects of the construction of the high-speed TGV Est railway line.

Again, it is the strategic planning documents of Saarland and Luxembourg that reflect most accurately the analytical vision of polycentrism. According to this, the Greater Region should be based on a collection of differentiated city networks with distinct functions: a network of major cities with specific and complementary functions, networks of smaller cities located on the peripheries of these and encouraging the development of decentralised links (such as the cross-border agglomeration of the European Development Pole [Pôle européen de développement, PED]) and, finally, networks of small cities within the areas less dense in terms of population and facilities.

It must, however, be noted that this polycentric vision of the interregional project is struggling to become established. The other parties to the Greater Region (Wallonia, Rhineland-Palatinate) that are more remote treat the Greater Region as only one axis of development among others, while the Lorraine region treats the "Lorraine corridor" as the key logistical axis for metropolitan development. Saarland and Luxembourg, which are the two smallest areas, need to anchor their spatial development within the cross-border environment for reasons relating both to land and economics. They are effectively the real centres providing attractiveness and driving forward cross-border relations.

Another obstacle is associated with the competition between states, regions and cities within the Greater Region: the latter is primarily a political tool available to regional officials who have shown little interest to date in bringing together the local actors who could ensure the effective implementation of certain recommendations and initiatives. While the cities of the Saar-Lor-Lux-Rhine area have been members of the Luxembourgish association EuRegio since 1995, to defend their interests, they have never cooperated on the preparation and development of strategic approaches for the various summits. The absence of a specific budget within the Greater Region programme (2007-2013) to support the formation of city networks is in contradiction to the other efforts made at the inter-regional level. The socio-economic and territorial imbalances generated by the exploitation of cross-border differentials should also be better taken into account in the development of these networks: while Luxembourg and, to a lesser extent, Saarbrücken can be considered as the real supporters of a metropolitan dynamic, the risk is high that other border districts, such as Forbach, Longwy, Thionville and Metz, will end up as no more than dormitory suburbs.

City networks: the real cross-border actors?

The QuattroPole and LELA+ city networks, which are fairly closely linked, represent above all forms of voluntary cooperation between local authorities (Map 9.2). By cooperating in this way, the cities are attempting to overcome their respective handicaps and affirm their positions within the Greater Region, which has largely ignored them.

Map 9.2 – The QuattroPole and LELA+ city networks

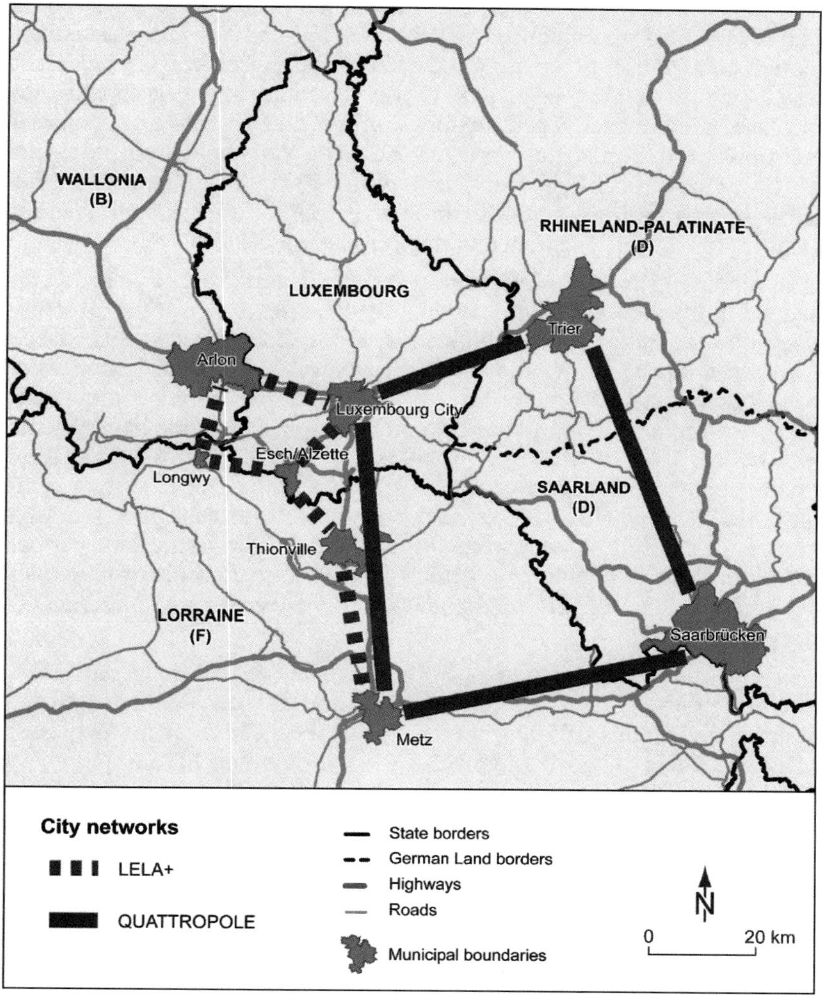

Author: Sohn, 2010.

QuattroPole, a polycentric virtual metropolis between Paris and Berlin

The four mayors who founded the QuattroPole network – Mr Rausch, the Mayor of Metz; Mr Helminger, the Mayor of Luxembourg; Mr Schroer, the Mayor of Trier; and Ms Britz, the Mayor of Saarbrücken – realised that after German reunification the centre of gravity of Europe had moved eastwards and thus, if stronger synergies were not developed with their surrounding areas, the four cities would find themselves marginalised. They thus decided to strengthen their mutual cooperation by creating a city network on February 29, 2000. It was noted from the outset that it was essential to work together to increase their attractiveness within Europe and to make this virtual metropolis into a major urban centre between Paris and Berlin in terms of communications and services. With 500,000 inhabitants, three airports, four railway stations, three river ports and four universities, QuattroPole is well-positioned at the European level.

Functioning in a decentralised way, the QuattroPole project is embedded in the administration of each city. The flexibility of the structure allows collaborations to be initiated in a pragmatic way in accordance with the interests of the member cities. Several working groups were constituted at the start in the areas of infrastructure and telecommunications, to implement high-speed communications network between the four partner cities (Quattronet in 2002), resulting in the implementation by the operators in charge of the project of a unique tariff system dealing with the legislative differences between countries. Other working groups were set up to encourage tourism and culture in the four partner cities, making available to visitors a joint tourism infrastructure, or allowing the languages of the neighbours to be learnt via the internet (e-learning method).

Deliberately orientated towards the north and east of the *département* of Moselle, the QuattroPole city network initially constituted, until the municipal elections of 2008, the geopolitical objective of the mayor of Metz. A native of Sarreguemines, Jean-Marie Rausch can speak and understand German, which allowed him not only to strengthen his relations with the mayor of Saarbrücken but also at the national and regional level, via the various functions he fulfilled, to implement various projects of cross-border cooperation. Interested in new technologies, he made these into the spearhead of the economic recovery of the city and its change of image: in 1978 Metz became the first city in France to have cable; the first to have an internet site, in 1995; and since the 1980s it has attracted small and medium-sized businesses specialising in IT and new technologies to its science park, forming partnerships with universities in neighbouring countries and America. By positioning Metz as the "epicentre of a Luxembourg-Saarbrücken-Nancy triangle", the major of Metz has contributed to both countering the ambitions of Nancy and reinforcing a strategy of openness to the north of the *département*, followed since the 1970s.

The municipal election of 2008 appeared to render possible a strategic revival: for the first time since the end of the Second World War, the new mayors Dominique Gros in Metz and Bertrand Mertz in Thionville were of the same party (Socialist), and intended to create metropolitan links capable of bringing the two cities together and creating a continuous and attractive North Lorraine axis with 600,000 inhabitants, focused on the Luxembourgish dynamic, which is the strongest (INSEE 2010). The two mayors agreed to take up the preliminary project of a law on territorial reform, which proposed merging within the major "urban zones" the existing local authorities and granting them all powers associated with urban life.

> When one looks at the major cities referred to in the Balladur report, one sees that for the average French person, there is nothing between Lille and Strasbourg. Our territory is completely forgotten, as it lacks representation. We had to initiate, with all the relevant mayors and inter-district presidents, a process of polycentric metropolisation of the Metz-Thionville axis[2].
>
> The morphological continuity between the agglomerations of Metz and Thionville allows decisions to be taken at the level of 600,000 inhabitants. This dynamic is represented in particular by the flow of commuters who, while being reduced since 1999 between Thionville and Metz, has increased much faster in the other direction thanks to the Luxembourgish dynamic – the attractiveness to the north and the crossing of the borders are now more characteristic of the restructuring of regional dynamics than the "corridor" effect[3].

However, far from having the unanimous support of the grouping of 13 *communes* that the Metz-Thionville axis is intended to bring together, the latter has been subject to political wrangling within both the agglomeration of Metz (Communauté d'agglomération de Metz Métropole, CA2M) and the agglomeration of Thionville (Communauté d'agglomération Portes de France-Thionville), which the mayors of the two main cities do not control.

LELA+ develops its network on the basis of the Luxembourgish dynamic

Since the memorandum of understanding signed in Longwy on October 18, 2006, a new cross-border cooperation network has been created: the LELA+ network, which initially included the cities of Luxembourg, Esch-sur-Alzette, Longwy and Arlon, before expanding to include the cities of Metz and Thionville in September 2007.

The intention of its founders was that the LELA+ network would allow "flexibility and efficiency" to be combined. The aim was to take

2 Interview of Bertrand Mertz for the weekly newspaper, *La Semaine*, 2009/03/12.

3 Interview of Louis-François Reitz, Director of the prospective mission and metropolitan stakes at the municipality of Metz, 2009/05/19.

into account, support and also shape the main cross-border dynamics within an employment area in which, for the municipalities close to the border, over 40% of workers are employed in Luxembourg. This dynamic is, however, developing in an increasingly expansive manner, moving towards the interior of the Lorraine territory; while Thionville has 25% of its workers employed in Luxembourg, for the Metz agglomeration the figure is around 5% (12,000 and 4,000 workers respectively). According to the mayor of Luxembourg City, Paul Helminger: "The big idea is to become equal partners, sharing our experiences and our resources. The network has concrete contents with issues developed relating to the population."

Taking into account the importance of cross-border working in the region (see Decoville and Sohn, Chapter 11), the LELA+ network makes the issue of mobility one of its key aims at two levels of spatial analysis. On the one hand it seeks to promote the integration of cities into the major European infrastructure networks (road, rail, air) and, on the other hand, to promote accessibility between cities. This latter is achieved both by improving public transport between urban locations at the level of city networks, and within cities themselves. Thus one can view the establishment of such a city network as allowing such synergies between the nodes of the network to be engaged as to overcome the blockages created by the presence of borders, both national and departmental (in particular between the départements of Moselle and Meurthe-et-Moselle).

Improved digital accessibility is one of the other priorities of the LELA+ network in order to connect the university centres of Lorraine and the future Luxembourgish university centre of Esch-Belval. The cities of Audun-le-Tiche and Esch-sur-Alzette have thus increased their corporation in order to facilitate the implementation of a fibre-optic network, which now extends to Longwy and Arlon. Ultimately this cross-border operation should allow high-speed networking for IT, research and higher education applications between the six cities in the LELA+ network.

What is the role of the city networks in terms of the cross-border governance of the Greater Region?

In this final section we examine the role these city networks play within the Greater Region, and in particular analyse their ability to structure cross-border territorial governance. To do this we examine the advantages and disadvantages of this form of coordination of public policies.

Networked governance with the advantages of flexibility and pragmatism

The network structure has a certain number of advantages: relatively flexible to the extent that it is integrated into the functioning of the local administration, it is not accountable to other institutions of the Greater Region. For the LELA+ network, a steering committee formed by the six mayors of the partner cities is supported by a coordination committee responsible in strategic terms for preparing political decisions and administering the various working groups on the basis of the issues addressed. Decisions are then taken unanimously by the steering committee.

The two networks function according to the principle of a participatory budget, with an allocation of €350,000 for QuattroPole and €120,000 for LELA+ (2008). The budget allocated to the functioning of the city networks is taken directly from the municipal budget, without constituting a new burden for the municipalities. In addition, the amounts allocated limit the scope of investments without additional subsidies, obtained for example within the framework of Interreg. The issues addressed reflect the will to develop a metropolitan strategy. The latter is intended to allow the accessibility of the various hubs of the network to be improved (for example the improvement of cross-border mobility and transportation; high-speed communications), or their attractiveness (by intensifying urban renewal activities and touristic and cultural development). The various networks also develop their image in terms of territorial marketing and seek to promote a shared cross-border identity. According to Bertrand Mertz, the Mayor of Thionville: "We are entering into a Schumanian process, a policy of small steps to remove obstacles to meet the real needs of cross-border commuters."

The establishment of city networks can thus contribute to the establishment of a new form of coordination and territorial development at the interregional level. The network structure remains relatively informal but allows pragmatic cooperation to be initiated and the elements of a shared strategy to be implemented: QuattroPole and LELA+ are above all networks of local actors who, while exploiting borders, also create new territorialities, driving forward certain projects that can only be implemented at the regional or interregional level. The four cities of the QuattroPole network have acted jointly with the regional authorities and those of the Greater Region to benefit from improved rail services linking four urban centres. The same type of initiative has been adopted by the partner cities of the LELA+ network to encourage the relevant authorities to achieve traffic decongestion on the motorways toward Luxembourg at the Zoufftgen border crossing, which is a real bottleneck.

With sufficient attractiveness the QuatrroPole and LELA+ cross-border networks can also contribute to grafting regional urban subsystems onto the metropolitan structure, by connecting Thionville, Arlon, Sarreguemines and Forbach, which are already linked by cooperative structures, and grafting the transport infrastructure to one or more of the partner cities. The city of Trier has participated in the implementation of the Quadriga project, together with the cities of Wittlich, Hermeskeil and Bitburg, to promote action in relation to transport, tourism and health, economic assistance and the improvement of the built environment. For Trier, the participation in the QuattroPole network can benefit from this experience and strengthen its role as a river logistics centre.

The fact that Metz has, since the signing of the charter of objectives in 2003, combined the network of Lorraine corridor cities with Nancy, Thionville and Epinal can also contribute to promoting the Moselle valley as a European transport corridor and promote inter-university cooperation. The participation of Saarbrücken in the Saarbrücken-Moselle-Est cross-border agglomeration promotes cooperation at the local and cross-border level in relation to the regeneration of industrial sites and economic development with Forbach (Eurozone) and the improvement of public transport links (a train-tram has linked the cities of Saarbrücken and Sarreguemines since 1997). The participation of the cities of Arlon, Longwy and Esch-sur-Alzette is in part due to the same reasons: centres that provide (Arlon and Longwy) or receive (Esch-sur-Alzette) flows of cross-border commuters have every interest in cooperating between themselves and with the regional metropolises in order to establish proposals relating to the realisation of an integrated transport system for the area. Located at the margins of their national territories, these cities exploit the networks in order to avoid being once again marginalised by the process of redevelopment taking place within northern Lorraine. In this way, the PED cross-border agglomeration is intended to anticipate the movement of flows of cross-border commuters towards the Alzette valley within the framework of the development of the Esch/Belval project by 2015.

Even so, the reticular organisation in no way guarantees improved territorial coherence

The networking of cities and regional metropolises is only a first step in the establishment of a true cross-border metropolitan region. One aspect that is crucial to guarantee the effectiveness of this reticular organisation is the integration of interstitial spaces into the dynamic of the whole: these consist primarily of urban areas that, more than the effects of agglomerations, constitute the real area of influence of the metropolises. While the city, as the node of the network, can act as an intermediary vis-à-vis the inter-municipal structure to which it belongs to promote a snowball effect, the latter none the less remains more limited than if the whole of the area of cooperation were involved. In Lorraine, the participation of Metz and Thionville in cross-border networks only partially compensates for this limited territorial anchorage. Meanwhile, Luxembourg City suffers from the absence of an intercommunal structure. With 88,586 inhabitants in 2009, 138,329 in the agglomeration and 43% of total employment (STATEC 2009c), the capital concentrates within itself metropolitan functions that it exercises not only at the level of the whole country but also within the cross-border functional urban region (see Decoville and Sohn, Chapter 4). Despite this dominant position within the national urban network and the phenomena of peri-urbanisation, which the city generates within its near and more distant environment, Luxembourg City does not enjoy an extended area of influence. In a small country, the local authorities constitute, particularly in rural areas, a strategic level of territorial anchorage for lobbying at the national level to the Chamber of Deputies.

There are also possible effects of competition between city networks. The Nancy-Metz rivalry has resulted in the development of separate city networks, and has anchored the Lorraine corridor in a regional dynamic limited by the border, as the metropolitan cooperation project it created has not involved Luxembourg City. The effects of competition can also come into play between the major regional metropolises, in particular Luxembourg and Saarbrücken, to reinforce their positions as cultural centres.

These city networks cannot truly take on a critical role unless they manage to combine within a metropolitan dynamic both urban and rural regions, with this split being particularly marked within the Greater Region. While an urbanised and industrialised Saar-Lor-Lux developed around the cross-border conurbations and agglomerations or city networks, the issue arises from the survival of a much more rural Saar-Lor-Lux, where population densities can be below 15 persons per km^2, which is among the least dense areas in Europe. Within the Greater Region, this space borders the interregional area on its western edge from the south of Wallonia and Meuse to the Vosges mountains and the Ardennes forest on its eastern edge. In order to meet this objective of territorial cohesion, the executives of the Greater Region have decided to promote a new form of polycentrism, using cross-border national parks[4] as the main instrument of a policy of local development and

4 Speech given by Jean-Marie Halsdorf, Luxembourgish Minister for Home Affairs and Town and Country Planning, on 17/02/2009 on the occasion of the international conference entitled "The national parks of the Greater Region: drivers of regional

environmental preservation. Just as city networks are intended to allow the progressive structuring of a process of metropolisation, the national parks are intended to ensure the synergetic interaction of "local, regional and cross-border actors, allowing the establishment of joint projects in different spheres, such as the protection of water resources, the promotion of natural heritage or the preservation and sharing of historical sites." The combination of these two processes has not yet really developed. The fact that the process of the peri-urbanisation of the major cities is expanding ever further into rural areas means that Luxembourg, Saarbrücken, Metz and Nancy are all faced with the question of how to deal with the phenomenon of urban sprawl.

Conclusion

The creation of the QuattroPole and LELA+ networks has undeniably forced regional and above all national elected representatives to take notice of the shift of regional dynamics towards the north of Lorraine and the border regions in general. The map of the distribution of wealth now shows the border regions of Luxembourg with average incomes among the highest in the region, rivalling those of the Metz agglomeration, while the southern Meurthe-et-Moselle is far lower (INSEE 2009b). The strengthening of the migratory movements towards the north of Lorraine, heavily within Luxembourg's sphere of influence, constitutes the driving force behind this, and also poses the question of the development of the so-called "residential" economy to meet the service needs of these new arrivals.

A cross-border metropolitan structure may be of assistance here, to the extent that it allows urban centres whose influence would otherwise be limited, to be joined and be visible to decision makers, to compete with larger European metropolises. The establishment of a polycentric urban framework can also allow certain intra-regional rivalries to be overcome; these exist at all levels between the territories of the Greater Region and within the various areas that constitute it.

Having developed independently and in parallel to the Greater Region, the QuattroPole and LELA+ city networks enjoy a certain degree of autonomy with regard to the definition and implementation of their objectives. However, because they rely on funding from municipal budgets, they do not have the political and financial support necessary to undertake major actions. It is therefore to be feared that in the absence of the transfer of certain powers, in particular in relation to regional development and urbanism, these new territorial structures may not

---

development of rural areas" within the context of the preparations for the 11th summit of the Greater Region.

become very effective. The definition of their sphere of influence appears to be equally crucial to counteracting the centrifugal dynamics towards the north and south of the Lorraine corridor, and to dealing with the risk of the increasing depopulation of large, peripheral urban areas, or, conversely, to better integrating peri-urban sectors facing urban pressure that is channelled to a greater or lesser extent, above all in the border areas. It is thus important for the larger urban centres to better integrate their surrounding areas in order to promote a development of the regional or interregional territory as equal as possible, otherwise there is no guarantee that this polycentric metropolis will be able to achieve the goal of greater territorial cohesion.

While the Greater Region may be in a position to consolidate its presence within a European framework, this objective remains to be achieved and demonstrated, in terms of both territorial and economic relevance and of its structure. It appears that the ambiguities found within the Lisbon strategy are replicated here: how can the long-term harmonious development of the European or regional areas be reconciled with the reinforcement of the supremacy of the major capitals or metropolises?

CHAPTER 10

The Challenges of the Spatial Development Policy in Luxembourg

Antoine DECOVILLE

The process of development of the main cities of Western Europe is heavily influenced by the confrontation between two contradictory dynamics. The first is the process of metropolisation, which takes the form of an ever-increasing spatial concentration of populations, activities and wealth within the urban agglomerations that are well interconnected at the global level (Ascher 1995, Halbert, Pain and Thierstein 2006). This dynamic is centripetal and creates inequalities in terms of spatial development since it relies on an economic system that favours territories and scales over others as sites for capital accumulation (Brenner and Theodore 2002). Toward this trend, a second dynamic, which is fostered by national regulatory policies, aims to implement spatial regulation measures. These initiatives have sought to encourage a more even distribution of people, activities, facilities and means of communications within space (Merlin and Choay 2000). The objectives are to reduce disparities between regions while maintaining a high level of economic development at the national level, and are based on the growth of regional urban centralities. In addition to their economic objective, these policies led by central states also include social considerations (such as the preservation of territorial cohesion) as well as ecological ones. Since the mid 1960s, which marked the "golden age" of spatial planning, these policies have often relied on a regional growth pole strategy, aimed at combating the negative externalities of uncontrolled spatial concentration around the most important cities by reinforcing the concentrated development of secondary centres (Parr 1999a, 1999b).

Policies of deconcentration have been adopted in centralised and federal countries, but with few successes, showing the relative inability of the public authorities to achieve this political objective of spatial development (Bontje 2001). Considering these limitations in terms of efficiency, and due to a less favourable economic context, the aims of the territorial policies have been significantly revised. The reduction in public expenditure, the process of decentralisation and the building of the European Union have contributed to the reorganisation of political power, particularly in terms of spatial regulation.

Today, this reorganisation is characterised by a less central position of the state in the decision-making process and by the adoption of neoliberal urban policies aimed at improving competitiveness in order to attract investors at a local level (Brenner 2003, Genestier, Jouve and Boino 2009). Spatial development has changed in terms of both its objectives and its means; it now often seeks to liberate the potentials of territories with the assistance, where applicable, of the state, rather than to reorganise people and activities in the space. These changes are also a sign of a redistribution of powers and a new way of thinking about the role of the political sphere in the management, development and regulation of the territory in which it is located. Nonetheless, thanks to its power to assign funds, the state still has the scope to act to promote the development of particular territories; it coordinates territorial strategies deployed at an increasingly centralised level and thus has not lost its means of influencing spatial development. It is able to participate in decision-making processes that relate to territorial projects by being involved in the new forms of governances developing at local levels.

In this context, the Luxembourgish case study presents specificities, linked to its size, its administrative structure and its political culture. The aim of this chapter is to present the particular way Luxembourg addresses the political issue of regulating its spatial development, in the context of a process of metropolisation in which economic dynamics appear to play a primary role.

The first section will be given over to the study of the evolution of national spatial planning policies in Europe. This part will allow for a better underlining of the specificities of Luxembourg, which will be addressed in the second part. In the third step, the major operation of urban development currently being carried out in Esch-Belval will be examined, as it offers a synthesis of the strategies that underpin the spatial planning policy enforced in Luxembourg.

Issues and changes in policies of spatial development in Europe

To understand the way in which spatial development has been conceived, formulated in regulatory documentation and then implemented in Luxembourg, it is necessary to briefly review the major stages in the evolution of spatial development policies in Europe. In this field, Luxembourg has drawn inspiration from both the theory and achievements from much larger countries, despite its specific territorial, economic and social characteristics.

Order, hierarchy, equilibrium

The major objective of Keynesian policies of spatial development implemented in Western Europe has been to reduce disequilibria between regions by reinforcing secondary centres. The work of Christaller on central locations (1933) was the best-known formalisation of this theorisation of the national urban system, aiming to optimise access of the whole of the population to services. This theoretical frame had a significant lasting impact on the spatial development policies adopted in Europe from the 1950s onwards. Whether in France, with its *métropoles d'équilibre*, the Netherlands, Italy, Germany, Switzerland or Denmark (Newman and Thornley 1996), the principles of territorial equilibrium have been taken up and refined on multiple occasions, in accordance, of course, with national contexts. From the 1950s to the 1970s, states deployed a whole range of regulatory strategies aimed at balancing economic growth in favour of less-developed regions and rural peripheries (Brenner 2004). The role of the state was to attempt to maintain an equilibrium of development and to encourage a better spatial distribution of wealth. For Brenner (2004), this era was the age of "spatial Keynesianism", which was characteristic of the period of strong post-war economic growth. According to Brenner and Theodore (2002: 355), these spatial planning policies had also direct economic reasons:

> Uneven development is associated not only with new opportunities for capital but also with any number of potentially destabilizing effects that may undermine the "structures coherence" upon which sustainable capital accumulation depends. In response to these persistent dilemmas, capital states have mobilized a variety of spatial policies intended to regulate the uneven development of capital.

These policies did not, however, enjoy the expected degree of success (Bontje 2001, Parr 1999a, 1999b), which tends to demonstrate that the centripetal forces of the spatial concentration of people and activities appear to prevail over the forces of dispersal, despite the improvement in various means of communication which could, prima facie, allow growth with less regard to spatial proximity (Cook 1991, Halbert, Pain and Thierstein 2006). In other words, the incentives or regulations put in place by spatial development policies in order to promote places identified as development priorities are often less important than the gains that can be earned by investors when they tend to concentrate. (Thisse and Van Ypersele 1999). The end of the period known as the "thirty glorious years" in the 1970s probably did not help to reverse this trend, as it forced the authorities to review their modes of action in terms of development in order to better deal with a context of increasing economic uncertainty. This resulted in a process of decentralisation, which led to the state playing a less central role in spatial development policy. In parallel, the local levels were reorganised, with the institutionalisation of forms of collective action within strategic urban planning (Le Galès 2004).

Less regulation, more collaboration

This neoliberal turn led to a fall of direct investments by the state, which encouraged cities to find alternative sources of funding by mobilising local economic actors. Market forces took the place of the regulating state (Weber 2008), and real competition between local territories developed, motivated by the need to capture investment (Kearns and Paddison 2000). This redefinition of the action of public authorities sometimes favoured closer proximity between the holder of political power (local elected representatives) and the holders of economic power (the business world), which made the realisation of major urban projects bringing about economic development possible. The urban policies known as post-Keynesian, designed locally and implemented by coalitions of public and private actors structured at the city level, thus replaced the Keynesian urban policies implemented by the state in order to meet objectives in terms of facilities and development (Dormois 2008). In this context, which puts territories in competition with each other, the metropolitan regions occupy an increasingly key role. They are the centres of endogenous economic development that can increase the competitiveness of the national economy (Krätke 2007). While this demonstrates the power of metropolitan regions, as territorial actors, to challenge the central and supreme position of the state in the sphere of decision-making, this is not done against the state but rather with the state's support, within a system of mutual dependency (Le Galès and Harding 1998). In order to preserve a certain leverage over the metropolitan regions, the states have moreover sought to guide their emergence at the political level, sometimes even by initiating the creation of metropolitan institutions, in the mould of the "urban communities" in France or "metropolitan cities" in Italy (Antier 2005).

In parallel to these developments in terms of powers, spatial planning policies have also witnessed a restructuring of their strategies, methods and means. Under the influence of the "competitiveness" agenda (Albrechts 2006a), they have gradually evolved from a land-use allocation approach towards a more strategic perspective, focused on the carrying out of projects (Motte 2007, Schmidt 2009). This approach is therefore focusing more on relations and processes than on space and place (Albrechts 2006b). These mutations seek to provide greater flexibility and responsiveness of public action in a context of uncertainty and slower economic growth (Motte 2007).

In this context the state has nonetheless not entirely renounced all intention of controlling and managing territorial dynamics, although the forms and means used have changed greatly, as is made abundantly clear by the literature on "state rescaling" (Brenner 2004, Jessop 2002). It tries to solve the paradox between two contradictory objectives. The objective of reducing territorial inequalities, initially linked to the idea of the state as an agent of development, regulation and intervention, as opposed to that of reinforcing the main urban centres, drivers of economic development for the whole country. The aim is to allow these major centres to attain the critical mass to find a place within the ultra-competitive global urban network, which generates an ever-increasing share of global wealth.

Faced with this paradox, the state is constrained to opt for compromise and to find new strategies and tools. The focus on polycentrism can be interpreted as an attempt to compose with these contradictions. This spatial vision involves promoting the emergence of dynamic regions based on strengthening the cooperation between cities that are expected to constitute, all together, polycentric metropolitan regions (Meijers 2005, Davoudi 2003, Shaw and Sykes 2004, Gloersen *et al.* 2007). The aim is to favour the development of urban centres within regions whose endogenous potential appears underexploited, while avoiding the counterproductive effects of hyper-concentration in the primate cities. This policy of reinforcing secondary centres is not aimed only at simply adding dynamism to the nodal points within the territory but rather is equally based on a supposed knock-on effect of metropolitan regions on their surrounding areas, allowing the objective of cohesion to be met. Despite the use of new terms and concepts, this policy of spatial development does not differ fundamentally, in its theoretical principles, from those pursued by European states for the last 50 years. The European Spatial Planning Observation Network (ESPON) report entitled "*Potentials for polycentric development in Europe*" went as far as to argue that the aim of reducing disparities between cities via the development of polycentrism constitutes a new version of the traditional regional policy (ESPON 2005). European development policy appears, however, to be confronted with a major dilemma, between choosing to favour the economic competitiveness of the continent, in accordance with the Lisbon strategy, and supporting territorial cohesion (Lüthi, Thierstein and Goebel 2008). The ESPON report distinguishes between the policy of reducing inequalities practised in 30 European countries and that of the development of urban competitiveness followed in 19 other countries. But must these two dimensions be in opposition? The difference probably resides more in the tone that the political actors wish to give to the discourse around their spatial development policy than in any real difference in the tools and methods used. Equalisation of the level of economic development can only be posed as an objective if this is to be achieved to the benefit of the less-developed areas and not at the expense of the better-performing ones.

Despite appearances, it would probably therefore be misleading to interpret the changes in the modes of action of states in matters of spatial development as revealing a major change of course in Europe. Rather, it appears that there is a will to adapt the tools and means to social, political and economic changes, while still pursuing the same goal: a more balanced urban system, providing both fairness and stability for the ruling powers.

Following this brief overview of the issues at stake and the transformations undergone by territorial policies in Europe, Luxembourg appears to be a very particular case study. This is a relatively atypical example, both in terms of the general principles that rule its spatial and economic development and the involvement of the state in the forms of governance structured around urban development projects.

The political support for spatial development in Luxembourg

The analysis of the spatial development policy in Luxembourg is based on the experiences gained from the work conducted for the Ministry of the Interior and Spatial Planning, (CEPS/INSTEAD 2008) as well as meetings with policy experts and decision makers. The purpose of this part is to present how the state attempts to regulate the spatial development of the country confronted with a process of metropolisation and to identify its deadlocks.

A recent policy concern for a specific territorial configuration

Spatial development policy in Luxembourg has specific characteristics linked to the national context, in particular the small size of the territory (2,586 km²). As a result, and due to the significance of the phenomenon of metropolisation, it is not possible to identify a separate regional subsystem with its own dynamics of spatial development. On the contrary, the metropolitan area of Luxembourg City not only covers the whole of the country but also extends to a great degree beyond the national borders (see Decoville and Sohn, Chapter 4). This process of metropolisation is mainly related to the economic dynamism of the financial place, the relative importance of employment relating to the knowledge economy (Sohn and Walther 2008) and the very high level of attraction exercised by the capital city over the border regions.

Concerning the institutional structure in place in Luxembourg, there are only two different political levels: the state and the communes. One should nevertheless mention that an administrative reform is currently under discussion, which could lead to the creation of urban communities and encourage the fusions of communes in the future. Despite these specific characteristics, the principles and tools used to support this policy are surprisingly similar to those adopted in more classical territorial and politico-institutional configurations.

In this context, marked by strong economic growth that is highly polarised, the adoption of a policy that aims to regulate the process of spatial development is relatively recent. The first legislation relating to this issue dates from March 23, 1974. This led, in 1978, to the implementation of the first "*Programme Directeur d'Aménagement du territoire*" (MIAT 2003), which already took a relatively comprehensive approach towards the challenges the country faces. This programme presented a systematic vision integrating the issues of urban development, economic growth, environmental conservation and cultural and natural heritage. The programme, however, was rapidly overtaken by the magnitude of the economic and demographic growth, which led to significant pressure on the territory.

A highly classical theoretical foundation

A new law concerning spatial development was passed in 1999, which expressed the need for a more economical use of land in the country. This law led to the adoption of a second *Programme Directeur d'Aménagement du territoire* in 2003, which established the framework for spatial development still in force. This programme proposed a subdivision of the national territory into six development regions, in order to encourage the communes to cooperate more by grouping them together for the purposes of developing regional masterplans (map 10.1). Today, this regional division is struggling to become reality, and only the South region of Luxembourg has an office conducting territorial studies and marketing. A pre-operational tool supports these general principles of the *Programme Directeur*, the Integrated Transport and Spatial Development Concept ("*Integratives Verkehrs- und Landesentwicklungskonzept*", or IVL). This operational interpretation of spatial development specifies the tools that will be created in order to better regulate the spatial development. These tools will be the sectoral plans, currently being developed or adopted (Innenministerium 2004). These sectoral plans will, in the long run, ensure compliance with the outlines set out in the *Programme Directeur*.

All these tools are focused on the necessity of redeploying the dynamics of growth from the centre towards other focal points, in accordance with a strategy of "deconcentrated concentration" (Ministry of the Interior and Spatial Planning, 2003). However, recent developments show a persistent trend towards greater spatial concentration of activities in the region of the capital (see Walther, Chapter 1; Walther and Dautel 2010), and a delay in residential urbanisation in the suburbs further away from the centre (CEPS/INSTEAD 2008).

The 15 urban centres identified as the places intended in the long run to counterbalance the dynamic of concentration in the capital are known as "centres of development and attractiveness", with three different hierarchical levels. The capital constitutes a higher-level centre, as it is home to relatively rare and specialised services that cannot be found anywhere else in Luxembourg. Esch-sur-Alzette, in the south, and the Ettelbruck-Diekirch conurbation (called the "*Nordstad*"), in the north, are described as being mid-level centres, providing access to services that are relatively rare outside of the capital. This secondary level of the urban system is the one on which the national policy is focusing its efforts, since Esch-sur-Alzette (with the Belval project) and the Nordstad are the only real places in the country that can assume to counterbalance the actual process of metropolisation, which induces a concentration of population and activities in Luxembourg City. But so far, the development of the Nordstad is far from being as advanced as Esch-Belval.

Map 10.1 – Luxembourg from the point of view of spatial development: regional divisions and central places

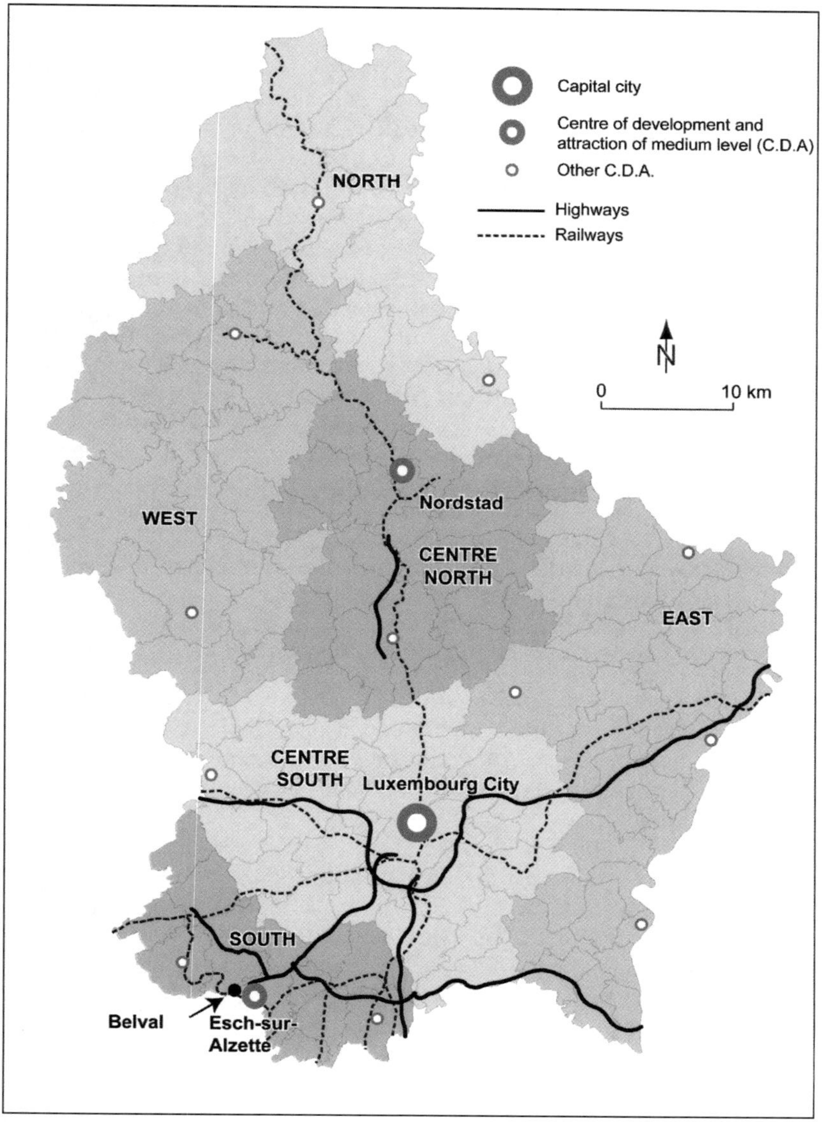

Sources: Administration du Cadastre et de la Topographie, Direction de l'Aménagement du territoire et de l'Intérieur – Ministère de l'Intérieur et de l'Aménagement du territoire. Author: Antoine Decoville, 2009.

This vision of the optimal territorial organisation to be pursued within the framework of the spatial planning policy is ultimately rather classical, inspired by Christaller's model (Christaller 1933) and with the guiding political role being played primarily by the state. Such a situation is easily explained by the fact that the state is the only political-institutional level above the communes, and therefore it is currently also the only one that can develop a strategic reflection on a supra-communal level. Moreover, the government of Luxembourg not only sets the broad principles and guidelines that have to be followed by the communes, but it is also the instigator of very concrete projects of urban development. As a result, the state is active on several scales, from the creation of the legal framework of spatial planning to the implementation of concrete urban projects, where it assumes its leadership role in the urban governance. The next part will describe the example of Esch-Belval, which appears to be the most important urban development project implemented by the State in order to fulfil its spatial planning objectives.

The Belval project, the major element of the national spatial planning policy

The urban development project for Esch-Belval consists of revitalising the former industrial brownfields in the communes of Esch-sur-Alzette and Sanem, in the south of the country.

Beyond the brownfields, a highly strategic area

This project, covering an area of 120 hectares, is expected to create around 20,000 jobs in the long run and lead to the attraction of 5,000 new inhabitants by 2020. Although there is so far no evidence that the project will fulfil the expectations of the state, the mobilisation of important public funds, estimated at over one billion euros, demonstrates a strong political will. This project fits in with the desire of the state to rebalance the territorial organisation of the country. Esch-sur-Alzette is the country's second city in terms of number of inhabitants, and is situated less than 20 km away from the capital. It thus constitutes a secondary urban centre with a real potential to redistribute part of the growth that is, so far, concentrated to a very great extent in the agglomeration of Luxembourg City. The objective of developing this district is all the more delicate as the city of Esch-sur-Alzette has a certain vulnerability in socio-economic terms (like the other districts in the southern region), with an unemployment rate of almost 10%, in contrast to 5.4% for the remainder of the country in 2008. The development of Belval is thus seen by the communes concerned, as well as the Luxembourgish government, as an opportunity to rebalance the local employment market and escape from a situation of heavy dependence on the capital.

So far most of the establishments on the site are public initiatives, such as administrative buildings, public research centres, university premises, and even a public concert hall. Concerning the private investments, the results are more discreet for now, with the exception of Dexia Bank, which has benefited from attractive conditions to settle in an impressive red tower, which embodies the renewal of the whole site.

A public/private coalition behind the project

The first foundations of close cooperation between the Luxembourg state and private investors in connection with the regeneration of the Belval site were laid in 1996, at a three-party meeting between the state, the social partners and Arbed (a steel company that merged in 2002 to form the Arcelor group, which itself merged in 2006 to become ArcelorMittal). This meeting was not unusual, as the steel group had acquired such power in the Luxembourg economy, since its creation in 1911, that the political representatives had included it in discussions about the country's economic issues (Bouquet and Ousey 2008). Also, in 1996, an ad hoc economic interest group was set up and produced a report stating the particular issues at stake with the Belval site, due to its significant size and strategic location. An initial strategy focusing on the diversification of the activities in order to consolidate the economic fabric of the southern region was then produced in 1999 (Del Biondo 2009). This initiative was also seen by the steel group as an opportunity to increase the value of its land, which was degraded by decades of industrial activity. The promotion of this urban development project was thus an aim shared to a large extent by the two main stakeholders. This convergence of interest was formalised with the creation in October 2002 of a private company called Agora, whose shares are divided equally between the state and ArcelorMittal. This company's aim is to manage the promotion, development and sale of the areas intended for private investment.

In parallel to Agora, a public body was created on July 25, 2002 to manage the state investment in the area. The *Belval Fund* is responsible for the building of numerous public facilities, including the future premises of an important part of the University of Luxembourg. The hope is to encourage private actors to follow the lead of the state in investing in an area that benefits from the proximity of public research institutions, such as the public research centres and the university. This division between the property development company Agora and the Belval Fund allows a clear separation of public and private investment, and thus avoidance of the project appearing too much as a political project, which could discourage private investors. This strategy is reflected in the public relation policy and in the name of the company,

Agora, which explicitly refers to a willingness to exchange and dialogue. The state thus plays a discreet role in terms of its involvement in the project, although it remains the major initiator of it. The Belval project is thus clearly distinct from the urban development operations in the Kirchberg plateau, which is older and home to the European institutions as well as many financial companies, in Luxembourg City. The urbanisation of this plateau, which began in the 1960s, was guided entirely by the state, which acquired land via a specific fund, including via expropriation (Helweg-Nottrot 2001). The marked differences between these two projects in terms of the method used and the clearly visible involvement of the state are a sign of the change in approach.

A weak involvement of the communes in the building of a territorial vision

In comparison to the state, the local public actors have relatively limited decision-making powers. The communes of Esch-sur-Alzette and Sanem have minority positions on the board of management of Agora with two votes, as against double this number held by the state and by Arbed. Nonetheless, the presence of these communes within the control structure reduces the institutional distance between the administration and the private developers, and thus simplifies the process of approval of planning documents at the local level. The communes are in any case very favourable to the project, which will provide an economic boost, even though their scope for manoeuvre within the implementation process is limited.

Furthermore, the establishment of a European Grouping of Territorial Cooperation (EGTC) is underway, which will give the project a cross-border scope. This type of European tool is intended to facilitate cooperation between border territories, and this is essential for the success of the whole project. Indeed, the French stakeholders have decided to accompany the project through the development of a cross-border eco-city, which will be built with the support of the French State. It shows that if at the starting point of the project the cross-border dimension was not so much taken into consideration, it has become a very important issue, and both French and Luxembourgish stakeholders have a lot to win in cooperating around this project. This EGTC will bring together local and national French and Luxembourg actors, but given the low level of funding and human resources available to the local authorities, it is likely that the states will remain central to the decision-making process.

The Belval project appears to be a great opportunity for the whole south Region. However, it is interesting to see what can be the possible consequences, in the mid and long-term, for Luxembourg City. Indeed, the capital is witnessing on its periphery the development of a new urban centre, which will attract activities and wealth. Moreover, the relocation of the University to Belval will keep Luxembourg from having an international academic recognition. So how will this spatial development project redistribute the balance of power between the various components of political power, and how is Luxembourg City, the centre of this metropolitan area, positioning itself?

Although Luxembourg City dominates the urban system of the country to a large extent, it remains a metropolis under construction and does not meet all the usual criteria that are usually expected from a metropolitan pole. The capacity of Luxembourg City to maintain its attractiveness at the global level will depend in the long run on increasing the range of services it is able to offer to both businesses and individuals. To date, the city does not have an economic promotion or strategic forecasting service, as the market has been largely favourable to the selective establishment of the organisations adding the greatest amount of value in the capital (Walther and Dautel, 2010). The choice made by the political elites appears to have been rather to promote locally the quality of life, in a context of increasing saturation of circulation pathways. Nonetheless, Luxembourg City will certainly develop its strategy in the future to play a more active role and to defend its interest if the economic situation appears to be less favourable.

The potentialities for increasing the efficiency of the spatial planning policy

Despite the undisputed leadership of the state in the field of spatial planning, the success of the policy of concentrated deconcentration remains very limited so far (CEPS/INSTEAD 2008).

A first reason can be found in the fact that the communes may not have always followed the recommendations of the IVL, and this may be due to the fact that they were not sufficiently aware or involved in the development phase of the spatial strategy. The lack of inter-municipal cooperation, as well as the lack of human resources at the local level, do not lead to the emergence of spatial strategies that are coherent at a supra-municipal level. Moreover, the communes do not necessarily adhere to the objectives mentioned in the "Programme Director" of spatial development, particularly when these orientations go against the demographic growth of some of them. The success of a planning policy is facilitated when a consensus with the majority of the actors has been achieved and when a leader is recognised and accepted by all the stakeholders.

Another limitation of the planning policy in Luxembourg, but also often elsewhere, lies in the fact that different ministries are pursuing their own goals, which are sometimes in contradiction to the spatial policy. Spatial planning is a political domain with a broad spectrum of action that requires coordination. It is not easy to defend its interests, particularly when major economic interests are at stake. Moreover, the legal tools of the national planning policy in Luxembourg are very flexible, which makes it difficult to prevent some projects that are contradictory to the planning policy interests. As a consequence, it will be crucial to observe a better respect of the planning interests in the future, which should be the case with the future "sectorial plans" that will be adopted in the fields of housing, industrial districts, natural areas and transportation.

Finally, better involvement of other players, such as business or civil society representatives, in the process of developing the spatial strategy would increase the chances of success. Indeed, investors are key players in ensuring the success of a spatial strategy vision. Through their investments, which are based on market studies, they create jobs and contribute to the development of a place. Involving these players higher up in the development of a spatial vision would enable policymakers to better understand the strategies of the investors and thus to have a more relevant knowledge of the real economic potentials of the territories.

Conclusion

Spatial development, as originally understood, aims to structure the development of the national territory in order to improve the cohesion and equilibrium of its different component parts. It concerns therefore a fundamentally regulatory position, which needs to be transversal and multiscalar in order to ensure optimal management of the territory's different resources. This concept of spatial development today is challenged by the imperatives that are imposed on it. It appears that the neoliberal turn of the 1980s gives priority to economic development over the regulation of the dynamics of spatial development. This change is manifested in the multiplication of local initiatives that aim to ensure the promotion of urban territories, while the planning of spatial development at the supra-communal level is increasingly flexible.

In this context, the example of Luxembourg appears atypical. The state remains the central actor of the planning policy, which defines the main principles on the one hand, but which also concretely implements its objectives on the other hand, particularly through the Belval project or, in the future, through the development of the Nordstad. In this way, the development of Belval appears to be the result of a classical vision of a spatial development policy with innovative processes in which public/private partnerships play an important role. The Luxembourgish state has a strong leadership in driving the spatial vision and organising the decision-making process, and it can also afford to implement it. How then to explain the significant differences that remain between the objectives that were defined and the real dynamics? The assumption that we defend is that a planning policy has to be shared and supported by the different actors who participate in the process of spatial development, whether they represent public institutions, private companies or the civil society. The success of the spatial planning policy in Luxembourg in the coming years will probably depend on its ability to raise the awareness of the stakeholders about its importance in terms of social, economic and environmental perspectives.

CHAPTER 11

Building a Cross-border Metropolitan Region

The Political Stakes of Scalar Restructuring

Christophe SOHN and Olivier WALTHER

Over the course of the last few decades, the intensification of the process of the globalisation of economic and cultural flows, the restructuring of the territory of states and the reinforcement of regional integration have led to a transformation of the functions traditionally attributed to the borders of the nation state (Anderson and O'Dowd 1999). In Europe, the border is now more of an interface that encourages contact and exchange than a territorial barrier, playing a structural role in the differentiation of political and economic systems and the establishment of distinct cultures and identities (O'Dowd 2003). The multiplication and intensification of transnational flows of capital, goods, labour and information have generated cooperation and institutions of governance between local and/or regional authorities on either side of borders (Perkmann 2003). This emergence of cross-border regions is closely linked to the process of European integration and is embedded within a multi-level strategy of institutionalisation (Hooghe 1996).

Following these changes, the relationship between city and border has been given new perspectives. Now, border regions are no longer necessarily seen as peripheral spaces cut off from the phenomena of urbanisation and economic globalisation (Herzog 1991, Saez, Leresche and Bassand 1997). On the contrary, in the particular case of cross-border metropolitan regions, the interface function of borders, which have become increasingly permeable, offers cities the possibility of taking advantage of cross-border differentials and benefiting from the positive effects that they offer to businesses and cross-border workers (Ratti 1994).

In the case of Luxembourg, the construction of a cross-border metropolitan region is examined through the analysis of the process of spatial integration: the level and opportunities of interaction between areas and the will to cooperate (De Boe, Grasland and Healy 1999). On the basis of the distinction between functional spaces and institutional territories, it is then possible to distinguish the functional dimension of integration from its institutional counterpart, or, in the terms of Perkmann (2007), *market-driven integration* from *policy-driven integration*. The former relates to socio-economic interactions, while the latter concerns the initiatives taken by political actors in matters of cross-border cooperation (Reitel 2007, Sohn and Walther 2009). The main hypothesis of the article is that the two dimensions of integration do not necessarily coincide, the construction of a cross-border metropolitan region being marked out by temporary optima that result from a balancing of the opening and closing of borders.

In the first section of this article, an exposition of certain theoretical elements relating to cross-border metropolitan integration will be given. On the basis of a constructivist approach, the article will show that the emergence of cross-border metropolitan regions is a dynamic process within which the issues relating to borders, in particular the economic or symbolic advantages that they can offer, are a central element in the cooperative strategies implemented by the actors. The second part of the article demonstrates the size of the division between the functional integration that prevails between Luxembourg and the border regions and the mode of cross-border cooperation practised within the Greater Region over the course of the past two decades. This difference between the two dimensions of spatial integration can be interpreted as the outcome of a mode of economic and social development that is based on a niche policy of exploitation and of the earning potential of borders on the part of the Luxembourg state. The third part of the article highlights the current limits of such a situation and provides some clarification of the interests and perspectives that underlie the recent implementation of cross-border cooperation.

Theoretical elements on cross-border metropolitan integration

In order to examine cross-border metropolitan integration in Luxembourg and, in particular, the role of the frontier context in this process, it is necessary to define the way in which the concept of cross-border metropolitan region is understood. This theoretical clarification then allows demonstration of how the borders are no longer seen solely as obstacles to socio-economic relations but also as new resources.

A constructivist approach to cross-border metropolitan regions

In contrast to the "nominalist" approaches that consider the existence of cross-border regions to be self-evident, and taking into account the real or potential existence of functional interactions or historical, cultural, economic or political affinities, this article emphasises a constructivist approach, which considers cross-border metropolitan regions as an historically-specific process of social construction (Perkmann 2003). In line with the permanent restructurings of spatial and territorial systems, new regions emerge, develop and, sometimes, disappear. This perspective, which is inspired by the work of Paasi (1986, 1991), sees the metropolitan regions as created via a complex of political, economic, cultural and administrative practices and discourses. Of course, in their discursive construction, the actors often seek to mobilise shared cultural, economic or ethnic resources in an attempt to legitimate their view. Furthermore, as a product of social action, the cross-border metropolitan region can become an instrument of power capable of structuring economic and cultural relations, or fields of governance. However, the existence of a functionally integrated space is not an indispensible precondition for the reinforcement of cooperative relations between actors. This is corroborated by empirical observations. Examples such as Copenhagen-Malmö (Øresund) and Vienna-Bratislava (Centrope) show that the establishment of a dynamic of institutional integration on a metropolitan scale need not necessarily be linked to strong manifestations of functional interdependences (MOT 2006, ESPON 2010). In addition, the case of Luxembourg, as set out in this article, underlines that despite the magnitude of functional relations with border territories, the coordination of policy is not carried out on a scale adequate to the matters at stake.

Ultimately, as underlined by Anderson and O'Dowd (1999), many other elements can be at the root of a process of integration, such as a shared history leading to shared cultural or ethnic characteristics, or, alternatively, the juxtaposition of asymmetrical political, economic or cultural systems that encourage the exploitation of differentials of salaries, prices and regulations across borders. These remarks lead us to argue that the matching of the spatial magnitude of functional relations and the form of cross-border cooperation is not necessarily considered as an optimum to be attained. In the cross-border context, certain gaps observed in terms of integration can be explained by strategic considerations linked to the presence of a border and the advantages this can offer to actors.

The border – resource or recognition factor?

The fact that borders can constitute an advantage and offer protection (and not only a barrier to flows) has been relatively little discussed in the literature to date (Anderson, O'Dowd and Wilson 2003, Van Houtum 2000). From the end of the 1970s, however, certain authors have shown that the cross-border regions can benefit from their particular situation in the context of the European project. Reacting to work theorising location and centres of growth, Hansen (1977: 12), for example, formulates the hypothesis that "a stable border, together with a relatively unimpeded international labour and capital mobility will, on balance, be more advantageous than disadvantageous to a border region". On the basis of the example of the Upper Rhine region, his work provides the foundation for a theory of border regions. It is true, admits Hansen, that the political instability linked to the proximity of the border induces a certain economic risk, which is reflected in the prices and disadvantages of border markets in relation to cities in more central locations within countries. It is also true that borders divide complementary market regions, which necessitates a doubling of institutions or services that could function at lower cost in a homogeneous market. However, as has been shown by several subsequent works, border regions present unique opportunities (Leresche and Saez 2002), in particular cross-border metropolitan regions that have been able to take advantage, since the 1980s, of the progressive erasure of borders in the European Union.

The advantages of border regions can be classified in accordance with whether the border setting is seen as a resource, in particular due to the differentials and potential revenues it implies, or as a factor leading to recognition within the process of cross-border territorialisation (Sohn 2010).

The border constitutes a resource when it allows the implementation of an economic "model" based on the comparative advantages of each of the states. Businesses here can undergo a certain functional specialisation, with the headquarters and production facilities being located in the country with the best advantages in terms of tax or labour costs, a possibility already discussed by Lösch (1940) in relation to Switzerland and Germany. Being located close to a border can also allow certain regions to benefit from more developed services in a neighbouring country (operas, theatres, large museums) or comparatively cheaper products (fuel, cigarettes, food). This well-known advantage of border regions generally leads to a strong asymmetry in terms of provision of services and products. The presence of a border also allows advantage to be taken of property resources, as in the cases of Geneva (INSEE-OCSTAT 2008), the border between the Netherlands and Germany (Van Houtum and Gielis 2006) and Luxembourg (Carpentier 2010), for example. Households wishing to acquire property are able to find in the neighbouring border area accommodation that is both more readily available and cheaper. Finally, the presence of a border also allows territorial actors engaging in cross-border projects to benefit from additional financial resources from the EU. Since the 1990s, EU-funded programmes established within the context of initiatives aimed at stimulating interregional cooperation (Interreg) have played a significant role in the creation of new cross-border cooperation bodies, both from a quantitative (support for the creation of new initiatives) and a qualitative (transformation of existing less structured cooperative arrangements into more institutionalised bodies) point of view.

The border can also constitute a recognition factor for cross-border metropolitan regions, which use this to position themselves vis-à-vis other metropolitan regions in the same country or abroad. In Basel, for example, cross-border projects such as the Trinational Eurodistrict Basel and metrobasel are associated with the necessity to open up internationally in order to maintain the competitiveness of the city, situated in a region of low demographic growth and subject to competition from Zurich (Reitel 2006). Basel is thus seeking to improve its visibility as an international centre of art and culture, adding a cross-border dimension to events organised in the region, such as the international architecture exhibition IBA Basel 2020 for example. The presence of a border also allows certain actors marginalised in the national systems to increase their autonomy. By means of cooperation and alliances transcending the institutional and territorial divisions, these territorial actors can engage in projects that go beyond their strict national limits. Lastly, the presence of a border offers the possibility of inventing new forms of governance, especially given the great flexibility of the legal and regulatory provisions that constitute the framework of cross-border cooperation. This allows the implementation of institutional structures that function on the principle of networks and not only on that of hierarchy (Blatter 2003).

Cross-border metropolitan integration: the Luxembourg paradox

Applied to the case of Luxembourg, the analysis of cross-border metropolitan integration shows that the magnitude of functional relations between Luxembourg and the border territories, especially via the phenomenon of cross-border working, is significantly greater than the scale of the cross-border cooperative projects established to date. This situation must be put into the context of the dynamic of the urban and economic growth that Luxembourg has experienced over recent decades.

Strong functional integration on the basis of the flourishing of metropolitan activities

The flourishing of Luxembourg's metropolitan functions began in the 1970s. In this era, marked by the start of the decline in the national steel industry, the Luxembourg state decided to develop a policy of niches on the basis of tax regulation attractive at the international level. This strategy resulted in the establishment of an increasing number of foreign banks, accompanied from the 1980s onwards by an industry of higher service industries specialising in financial services (Walther and Schulz, Chapter 2). The economic growth that resulted from this policy of niches was significantly greater than that seen in other European countries, and led to Luxembourg establishing itself as an international financial centre specialising in the administration of funds and private banking. These higher service activities were concentrated in the capital, Luxembourg City, which is now home to over one and a half times as many jobs as inhabitants. The city is also the home of certain European institutions, which, like the European Commission, the Court of Justice, the Court of Auditors and the European Investment Bank, contribute to the development of an exceptionally well-qualified labour pool at the regional level. The growth of the financial centre and the European activities combine to support the economic and political centrality of the city on the European level. Luxembourg City is now a city with significantly greater influence than one would expect from its population (94,000 in 2011).

From the 1980s onwards, the growth of economic activity in the city and the agglomeration of Luxembourg led to a massive increase in cross-border workers entering from France, Belgium and Germany. While there were just 12,000 cross-border workers in 1980, there were 148,300 in 2009 according to the Luxembourgish National Institute of Statistics and Economic Studies (Institut national de la statistique et des études économiques du Grand-Duché du Luxembourg, STATEC). The attractiveness of Luxembourg to cross-border workers is remarkable, both in absolute terms and as a proportion of the national workforce – 44% of the latter consists of persons from Luxembourg's neighbouring countries. The focusing of the peripheral areas on the Luxembourg agglomeration is exceptionally marked: the metropolitan area includes 930,000 inhabitants, almost double the total population of the country (ESPON 2010).

The absence of cross-border metropolitan cooperation

In contrast to the magnitude of the functional integration that extends across Luxembourg's borders, there is not as yet any structure of cross-border cooperation at the metropolitan level. At the interregional level, Luxembourg plays an active role within the Greater Region, a cooperation area created in 1970 on the basis of a union of the state of Saarland (Germany), the region of Lorraine (France) and Luxembourg, initially named Saar-Lor-Lux, which Rhineland-Palatinate (Germany) and Wallonia (Belgium) have since joined. Although it has not been transferred any powers from the national authorities, this structure offers Luxembourg an institutionalised framework for cooperation which puts it in a favourable position vis-à-vis other regional units that do not have the prerogatives of a sovereign state. We will see in the third section that it is on the basis of this Euroregion cooperative structure that the project to create a polycentric cross-border metropolitan region – in the pipeline for years, deep within the administration of the Greater Region – was officially announced in 2009.

At the local level, one of the most significant projects of cross-border cooperation is without a doubt the European Development Pole (Pôle européen de développement, PED). Following an initial phase initiated by Belgium, France and Luxembourg which aimed at encouraging the economic transformation of the steel-working area of Longwy-Rodange-Athus, the local authorities of the three countries began to develop a project of cross-border agglomeration at the start of the 1990s. However, after two decades of cooperation, the results are mixed and the cross-border agglomeration project is currently in need of a second wind. The relaunch of the PED appears even more fragile as a neighbouring initiative has overshadowed it. The ambitious territorial development project of a City of Sciences on the former industrial wasteland of Esch-Belval, in the southern region of the country, is accompanied by cross-border cooperation between the city of Esch-sur-Alzette and the French local authorities in the vicinity of the site. Although, as in the preceding example, the cooperation is limited to the local level, the establishment of a European Grouping of Territorial Cooperation (EGTC) – a tool of cooperation with legal status – is a sign of the desire for institutional integration.

In a different way, Luxembourg City has developed cross-border cooperation in the form of networks of cities, which have linked in one case Saarbrücken, Trier and Metz (QuattroPole), and in another Esch-sur-Alzette, Longwy, Arlon, Thionville and Metz (LELA+). Carried out "under the radar of state institutions", these weakly institutionalised initiatives are embryonic forms of cross-border cooperation at a metropolitan scale. However, it remains the case that the objectives pursued by the various networks of cities mentioned are not of a metropolitan scope: the QuattroPole network concerns primarily a territorial marketing tool, while LELA+ allows cities to work together on a limited number of cross-border issues.

Exploiting the earning potential of the border, or merely making do?

The absence of cross-border metropolitan cooperation in Luxembourg can be interpreted in various ways. Putting the emphasis on the advantages offered by the border setting, the situation can be analysed as a strategy of the state of Luxembourg to preserve its room for manoeuvre and, in particular, its grip on the potential income resulting from the border due the fiscal and regulatory differentials (Sohn and Walther 2009). As the Luxembourg economy depends heavily on cross-border workers, and that this economic attractiveness depends in part on a policy of niches and the retention of certain protections associated with state sovereignty – particularly the fiscal and regulatory advantages in the areas of VAT, tax on individuals, companies and domiciliation (Pieretti, Bourgain and Courtin 2007) – the preservation of these cross-border differentials is of cardinal importance to the country. Moreover, this exploitation of the border as an economic resource through the use of cross-border differentials is of particular relevance to Luxembourg; the country is able to benefit from often highly qualified labour without having to bear the costs of its social reproduction. Lastly, in the event of a cyclical or seasonal economic downturn, the border position allows workers to be laid off without this impacting on the national unemployment rate. Given these comparative advantages that flow from the exploitation of cross-border differentials, the establishment of cooperation at a metropolitan level, accompanied by the redefinition of the parameters of intervention and a reduction in certain disparities between the territories, could be seen as a threat. In order to attempt to counter the risk of a metropolitan area becoming the object of coordinated collective action, and thus escaping its control, the state of Luxembourg has, it appears, favoured other levels of intervention in an attempt to manage the problems associated with the presence of borders.

This explanation, which is based on a maximisation of the utility of cross-border differentials for Luxembourg, must, however, be qualified. There are other factors that can explain the unwillingness of Luxembourg to develop cross-border cooperation on a metropolitan scale. Firstly, for cross-border cooperation to be established it is necessary for the actors to be in a situation that renders this necessary or desirable. In Luxembourg, however, the development of the city as a European and global hub is a recent phenomenon, and public actors have become aware of its cross-border dimensions even more recently. In addition, it has been shown that the physical proximity of a border is a factor that favours the establishment of cross-border cooperation, as this is the only way to deal with certain problems (Sohn, Reitel and Walther 2009). However, in contrast to Basel or Geneva, two other cross-border metropolitan regions, the national borders are not directly adjacent to the Luxembourg capital, but rather are around 20 km away, which is enough to alter the perception of border issues by the central power. Secondly, the emergence of a will to cooperate is not necessarily sufficient. It is also necessary for this vision to be shared by the key actors able to generate a consensus. Given a state is above all a collective actor made up of agents driven by potentially divergent or contradictory interests, the emergence of such a consensus could prove slow and difficult. Finally, the establishment of cross-border cooperation implies that the foreign partners willing to cooperate have been identified. In the context of the Greater Region, the territorial and institutional complexity implied by the union of six "regions" from four countries results in the presence of a multiplicity of actors with different powers and divergent interests, which is far from making the development of solid and effective partnerships easy.

Ultimately, however, the divide between the degree of functional integration and the institutional responses is interpreted, it should be borne in mind that such a situation is not fixed but rather is a temporary state in a dynamic process of institutionalisation of a cross-border region. The most recent political developments regarding cross-border cooperation within the Greater Region give reason to believe that a change of strategy is currently underway.

The development of a cross-border polycentric metropolitan region

Within the framework of the Luxembourg presidency of the 11th Summit of the Greater Region (2008-2009), the Luxembourg government proposed the launch of a process of reflection on the development of a cross-border polycentric metropolitan region (CBPMR). Three objectives are emphasised: strengthening the internal cohesion of the Greater Region; developing its position as a model region of cross-border cooperation in Europe; and consolidating its socio-economic influence and attractiveness, not only on the European stage but also beyond (Greater Region 2009). Although the active dimension of the process undoubtedly marks a break with the previous positions, understanding the motivations at the root of the creation of a CBPMR requires the examination of the underlying interests and the spatial changes engendered by the process of cross-border interdependencies.

The need to take into account cross-border interdependencies

Fostered by a policy based on niches and the shrewd exploitation of cross-border differentials, Luxembourg's economic growth is accompanied by increasing interdependence with the neighbouring territories. On the one hand, Luxembourg benefits from being highly attractive to companies and workers. But as a significant proportion of its employees live in the surrounding areas, Luxembourg is strategically dependent on this flow of labour. On the other hand, the cross-border workers benefit from Luxembourg's advantages in terms of salaries and social security, which results in a significant flow of revenue towards the border regions, in the order of 1950 million Euros for Lorraine in 2006 (INSEE 2009b). These financial flows are all the more important as some of these regions suffer from economic deprivation linked to deindustrialisation. Taking account of the magnitude of the cross-border interdependencies, the advantages to Luxembourg and the border regions can be seen to be accompanied by potentially harmful drawbacks.

Given the high number of cross-border workers, it is above all problems of transport that stand out. The saturation of the routes between the agglomeration of Luxembourg and the border regions is now such that it might hamper the development of the employment area and thus, the economic competitiveness of the region. To this is added the necessity of developing use of public transport, in order to combat greenhouse gas emissions and, more broadly, to promote sustainable spatial development. With regard to the territorial dynamics, the border areas benefit from far superior availability of land and property than exists in Luxembourg. As a result, residential migration can be observed on the part of a few thousand Luxembourg households who are deciding to live abroad, while maintaining their jobs in their country of origin (Carpentier 2010). The high price of property also serves to slow the relocation to Luxembourg of new foreign workers, which tends automatically to aggravate the problems of home-work mobility. The differential in residential attractiveness contributes to a process of functional division with a metropolitan centre concentrating activities and employment, on the one hand, and peripheral areas relegated to dormitory towns, on the other. Added to this functional division are mechanisms of socio-residential segregation, which feed into the cross-border differentials and could in the long run threaten the social and territorial cohesion of the cross-border metropolitan region (see Lord and Gerber, Chapter 6).

Given these various problems that exceed the territorial limits of national sovereignty, the solutions cannot be provided by a single actor but rather require ambitious cross-border cooperation. One of the stumbling blocks is the institutional and political fragmentation of the cross-border territory. The presence of four countries, each with their own specific cultural, territorial and legal characteristics, renders difficult the coordination of public action. Given the functional interdependence of the territorial units that constitute the cross-border system, the institutional fragmentation creates uncertainty. How can public policy be coordinated and coherent spatial development planning take place in the absence of dialogue and exchange? How can the divergent expectations and interests of a multiplicity of actors be reconciled in the absence of effective coordination? The need for cooperation to provide responses to the problems that affect the cross-border region is expressed in a desire to collaborate in order to benefit from the positive effects of territorial marketing. The fact that the cross-border metropolitan region is considered as the new territory of reference in international competition between metropolitan areas does indeed constitute a positive spur towards cooperation (Lefèvre 2004). But the development of cross-border cooperation at the metropolitan scale, whether forced upon policy makers or actively chosen, constitutes a significant challenge.

The challenges of a political construction in progress

The construction of a cross-border polycentric metropolitan region involves the establishment of a structure of cooperation charged with the implementation of the project. Rather than involving itself in setting out the outlines of a framework of cooperation adequate to the dynamics of integration operating between Luxembourg and its border territories, it appears that the executives of the Greater Region have chosen to favour the existing institutional framework.

This decision provides undeniable advantages but also significant downsides. In terms of advantages, it can be said that by favouring the Greater Region as the institutional setting, the actors in charge of the issue are spared the laborious process of identifying a new framework, with all that this implies in terms of political interests and power strategies, especially in a territorial context as complex as that of the Greater Region. In theory, the mobilisation of an existing structure could render easier the implementation of the project and save time, all the more so as the idea of a polycentric metropolitan region at the heart of the Greater Region has been brought up, in various forms, since the 1990s (see Lamour and Clément, Chapter 8). In reality, however, the choice of the Greater Region is problematic in that its territorial boundaries far exceed the functional area integrated into the metropolitan area: as a comparison, the metropolitan region focused on Luxembourg (4,344 km^2) represents just 7% of the total area of the Greater Region (65,400 km^2). For regions such as Rhineland-Palatinate, Wallonia and, to a lesser extent, Lorraine, the part of their territory directly affected by the process of cross-border metropolitan integration with Luxembourg remains marginal, which is clearly of relevance to the attitude of the authorities responsible for the management of these regions. Although there are many Euroregions such as the Greater Region in Europe (Perkmann 2003), their involvement in cross-border cooperation on a metropolitan scale is limited. Most often, ad hoc structures of cooperation have been created and these are relatively well adjusted to the functional urban regions whose territorial development they seek to coordinate. One possible solution would be to establish within the Greater Region a more restricted framework for action, on the basis of which the project of constructing the CBPMR could take shape. The definition of an "operational" Greater Region without these territorial margins would then be integrated into a multi-level governance approach, characteristic of European cross-border regions (Hooghe 1996). This solution would not, however, address the issue of its relation to existing structures of metropolitan cooperation such as the Saarbrücken-Moselle Est eurodistrict or the Sillon Lorrain network.

Beyond the question of the shape of the institutional structure, there is another issue concerning the actors involved in the cross-border metropolitan cooperation in the making. To date, the Greater Region has essentially involved the ten regional and state authorities that constitute the Summit of Executives. With a view to developing cross-border cooperation of a metropolitan nature, greater involvement of the main cities and municipal authorities appears essential. As nodes within the networks of globalisation and thus motors of economic, social and cultural development, the role of urban centres as collective and territorial actors must be recognised and integrated into the definition and conduct of the project. This is not a matter solely of consulting the municipalities and their local elected officials on the processes implemented by the state or regional actors, but rather of integrating them closely into the process of developing a vision and a common strategy. Besides considering the cities as political institutions, it is crucial to take into account economic and cultural actors, civil society organisations and more generally all the forces that constitute them in order to take into account the diversity of interests and give meaning to a structure that is at once political and social (Le Galès 2003).

Ultimately, the announcement of the creation of a CBPMR within the Greater Region does not resolve the issue of which geographical levels such a project should favour and the forms of cooperation that would flow from it. Although the executives of the Greater Region promote the vision of a polycentric metropolitan region, the dynamic of functional integration of the Luxembourg metropolitan area concerns a smaller area, organised around the Luxembourg capital as its central focus. Because polycentrism is a key factor in this multi-level institutionalisation of cross-border integration, close examination of the concept and its meanings is necessary in order to better determine the reasons for such a shift.

Polycentrism: territorial vision or political position?

The use of the concept of polycentrism presents a number of questions, which are not rendered any simpler by the diversity of approaches and definitions used to understand the phenomenon. As shown by

Kloosterman and Musterd (2001), Davoudi (2004) and Cattan (2007), the concept of polycentrism has a history similar to that of other concepts: initially developed as an analytical tool to aid in the description of reality, it has often been transformed into a normative agenda able to shape the direction of European programmes. If the creation of a cross-border polycentric metropolitan region were to become a priority, it would be nevertheless desirable to understand the polycentric character of the Greater Region, distinguishing between its morphological and functional aspects, on the one hand, and the different levels on which polycentrism can be studied, on the other.

Morphological polycentrism relates to the distribution of cities in space, while functional polycentrism combines descriptive criteria (what urban functions are involved?) and relational variables (what are the links between the cities?). It is essential to distinguish between these two dimensions of polycentrism in order to analyse the spatial organisation, as morphological polycentrism is often combined with functional monocentrism, as is the case in Brussels, London and the Netherlands (Hall and Pain 2006, ESPON 2007). It is also clear from studies conducted at the European level that polycentrism, where it exists, is essentially morphological. The level at which polycentrism is analysed is also seen to be determined by the functional dynamics and the adoption of suitable policies. Polycentrism indeed can be used to characterise the whole European level, as in the Spatial Development Perspective (European Commission 1999), the inter-urban level or the intra-urban level.

The analyses carried out to date suggest that the Greater Region is composed of several functional metropolitan spaces, which sometimes overlap and sometimes are discontinuous with each other (e-BIRD 2005, ESPON 2010). Certainly, all metropolitan areas concerned include several secondary centres, but do not have multiple centres in the way that Randstad does and none are cross-border except Luxembourg. The polycentrism within the Greater Region is thus above all morphological, which simply means that there are several urban centres of equal demographic size (Luxembourg, Metz, Nancy, Saarbrücken and Trier, to mention only those located in the heart of the region). The functional dimension of polycentrism, however, is much more problematic; analysis of the economic and urban dynamics shows rather a heavy focus on Luxembourg as the real economic engine at the heart of the Greater Region (Sohn and Walther 2008).

The polycentric dimension of the CBPMR in progress might thus be understood as a wish for a future that breaks from the present – assuming it is not merely political posturing aimed at reassuring the border partners of the "non-hegemonic" intentions of Luxembourg in relation to metropolitan cooperation. Whether or not this is the case, the question will be to what extent Luxembourg, as the dominant economic centre, will accept an equal distribution of roles within the mode of cooperation implemented. This remark leads us to ask, lastly, about the political compromises required as a foundation for the establishment of ambitious cross-border metropolitan cooperation. Without putting into question, to at least some extent, the economic, fiscal and regulatory differentials at the root of the functional asymmetries between the metropolitan centre and its border peripheries, the development of joint strategic plans leading to the construction of a CBPMR appears difficult. Considering the magnitude of the social and economic interests at stake in such a project, one can better appreciate the size of the task remaining and the magnitude of the challenge facing all the actors concerned.

Conclusion

In Luxembourg, the functional integration of the border territories into the metropolitan dynamic has not to date led to the creation of a political construction able to manage the development of a territory on a cross-border scale. However, rather than acting as a stumbling block, this failure has in fact sustained a remarkable economic expansion that has benefited both Luxembourg, as the centre of employment, and the surrounding regions via a redistribution of the revenues from cross-border workers. Taking into account the magnitude of the socio-economic interdependencies and the necessity of taking into account certain cross-border issues crucial to the survival of these territories, a change of strategy can be seen. Wanting to see the development of cross-border cooperation on the metropolitan scale, the Summit of Executives of the Greater Region now wishes to promote a cross-border polycentric metropolitan region. In the cross-border context, the emergence of new structures of cooperation, and more broadly the definition of new institutions, is a long-term process that results from the combination of macro factors (European integration, geopolitical changes, and globalisation) and micro factors relating to the power relations between actors.

While such a project will thus necessarily be long, complex and uncertain, the Luxembourg state is central to the issues. Having exercised its sovereign power and shown a certain skill in the exploitation of the border as an economic resource, the central state has now to enter into dialogue and hold negotiations with its border partners with a view to developing a new cross-border regime based no longer on asymmetrical functional relations, but rather on a shared and more balanced vision of economic and spatial development. The project of a cross-border metropolitan region may be a utopian one, but that does not make it any less necessary.

CONCLUSION

Luxembourg, the Emergence of a Paradoxical Metropolis

Bernard REITEL

When Roger Brunet published his work on European cities in 1989 (Brunet 1989), Luxembourg was not among the cases studied. A size criterion was used to select the urban agglomerations: Luxembourg, with less than 120,000 inhabitants, was under the threshold of 200,000 inhabitants. Less than 15 years later a new study (Rozenblat & Cicille 2003) used the same threshold but this time included Luxembourg City, while noting that this was an exception. This choice marked the recognition of an urban centre of modest size whose international and European significance is based on the presence of a financial centre and European institutions. To reduce Luxembourg's specificity to these three characteristics would be reductive, however; other elements must be taken into account. These are the cross-border dimension and the role of the state within the emergence of the metropolitan dimension. However, the most striking point is perhaps the speed with which Luxembourg has managed to establish itself and become recognised as a European metropolis: a centre of excellence on the global level (Lévy 1995), capable of driving growth across a huge region. As a result of this, four points should be noted. First of all, we consider the main characteristics of this metropolis. Then we examine the factors influencing the development of this metropolitan dimension by discussing the fundamental role of the state. Next we describe the issues with which this metropolis is confronted, while seeking to ensure its long-term survival within a European and global context. Finally, we seek to discover whether the characteristics displayed by Luxembourg are generalisable and reproducible.

Luxembourg: the strength of a small metropolis

Luxembourg's special character concerns above all the gap between its international dimension and its small size. The city's influence is significantly greater than would be predicted from its demographic weight. A small size is often viewed as a handicap as it results in less diversity in terms of resources and renders difficult the emergence of economies of scale (Bell and Jayne 2006). Meanwhile, studies have shown that small agglomerations have been able to exploit their positions within a competitive context by specialising in one or two economic niches and continuously innovating (Bell and Jayne 2006). The small size of an agglomeration necessitates the development of a strategy to make the best use of its resources. These are indisputable characteristics that apply to Luxembourg.

On the one hand, there is no doubt about the specialisation. The analyses show the existence of a hub for the management of mutual and investment funds. Finance is the main economic activity, which generates the greatest share of national wealth. It allows Luxembourg to position itself among the major urban centres of Europe, within a sector that is among those activities considered by the Globalization and World Cities Research Network to be decisive in their study of the process of metropolisation at the global level (Taylor 2004). This specialisation is relatively recent: it began in the 1960s within the framework of the economic restructuring initiated by the state.

A second type of specialisation is a little less recent and is linked to the European project. From 1950 onwards, Luxembourg was the head of the European Coal and Steel Community (ECSC). With the creation of the European Economic Community (EEC) in 1957, other European institutions were established there, such as the Court of Justice of the European Communities, the European Court of Auditors, Eurostat, Publications Office, the European Investment Bank, etc. The consolidation of the European Union and its successive enlargements have contributed to reinforcing this specialisation. Luxembourg thus developed into a legal centre on an international scale, but also into one of the three "capitals" of the Union. In this context, other specialisations have also developed such as pharmaceuticals, medical equipment and IT, but these activities, which provide a significant amount of jobs, do not yet appear to be of European significance.

On the other hand, Luxembourg has become a significant logistics centre. In terms of freight traffic, the airport occupies fifth place within Europe. The city benefits from its location within the backbone of Europe, but also from the presence of Findel Airport, and an airline, Luxair Cargo, which has managed to develop real expertise (Pansch 2009).

The growth of metropolitan employment has been vigorous for over two decades, which has allowed Luxembourg to consolidate its position within the system of European cities, and also within the Greater Region. The crisis of 2008-2009 does not appear to have put this into question, despite the temporary stagnation of employment.

In this context, the city's small size has more advantages than disadvantages. This makes it more likely that private and public actors who know each other will be in close proximity and establish informal relationships, which encourages rapid decision-making and allows rapid reactions to change. Furthermore, this small size strengthens the feeling of belonging to a small community. Being a metropolis is not simply a question of size but also one of strength, concentration and positioning. A metropolis is an "urban system whose configuration makes it possible to access to a level of excellence at the global scale" (Lévy 1995), and also a centre that seeks to establish a place in the world via an intensive network of relationships. Among the multiple characteristics that define a metropolis (Friedmann 1986, Leresche, Joye and Bassand 1995, Antier 2005), four dimensions appear for us to be essential and allow an overall concept to be defined: the presence of high added-value economic activities; a role as a node of exchange at the international level; innovative capacity and the establishment of a position within the knowledge society; symbolic power; and the fact of being a place recognised in the world (Knieling 2009).

Luxembourg does not meet all the conditions required to be classified as a metropolis, but whichever aspects are considered, significant changes have occurred over recent years. We have already seen that the urban agglomeration meets the first condition. In this context, the cross-border area is both a support for the metropolis, which provides it with some of the labour necessary for its development and a space over which the latter exerts its influence. Luxembourg has undeniably developed into a major urban centre between Brussels and Basel, between Paris and Frankfurt and between Cologne and Lyon. Certainly the second dimension, that of being a node of exchange, is open to discussion. Luxembourg enjoys good general accessibility due to its location along the European backbone, relatively close to global infrastructures, in particular those of Frankfurt and Paris (Bronner 2009). However, the city has only a modest airport and is not a major European rail hub. Luxembourg does, however, act as a gateway to the European continent for air freight, as the airport has cargo links to airports on the American and Asian continents. The role as an air hub concerns only goods traffic on the international level (Pansch 2009).

The analyses also emphasise the weakness in relation to the third dimension, that of innovation. Luxembourg is not a scientific centre on the European scale. The creation of the Public Research Centres in 1987 and of the University of Luxembourg in 2003 are still too recent for their effects to be felt, although progress has undeniably been rapid. In symbolic terms, Luxembourg has long suffered from a low degree of public awareness, and what does exist relates primarily to the institutional dimension. The organisation of the European Capital of Culture event, in 1995 and 2007, had a major impact (Sohn 2009a). On the one hand, the event allowed the city to acquire the cultural facilities of a true national capital: the Museum of Modern Art, the Philharmonic, etc. The buildings were designed by major contemporary architects; each building thus bears a signature that increases international visibility. Furthermore, the second hosting of the event allowed a cross-border dimension to be added thanks to the location of projects within regions of neighbouring countries. Public awareness of Luxembourg seems indisputably to be growing.

All in all, Luxembourg appears to be an emerging metropolis rather than an incomplete one. More than its form and dimensions, it is the process of metropolisation that is remarkable in Luxembourg, allowing the agglomeration to significantly increase its position over the past 20 years within the system of European cities. The international character is not only linked to the position within global economic networks, but also to the significance of Luxembourg's foreign population and its geographical proximity to three other European states. Due to these characteristics, multilingualism is an important advantage for Luxembourg in terms of international competitiveness. The emergence of this metropolis is, however, due at least as much to public as to private actors and in particular to the role of the Luxembourgish state.

A two-phase development

Before developing a reputation as a financial and European institutional centre, Luxembourg was seen rather as the archetype of a buffer state, located between two major powers, which placed it among the microstates. This position was especially awkward when the two powers, France and Germany, were in conflict. On the other hand, this was advantageous in times of peace. The metropolis emerged within the context of the European project and in particular the reconciliation between France and Germany, but was also linked to the desire of the Luxembourgish state to assert itself vis-à-vis its powerful neighbours.

In general, the growth of cities is linked to their capacity to develop activities by harnessing innovation, but also to the existence of a bourgeoisie with capital to invest in the means of production (Braudel 1979, Bairoch 1996). The presence of an active bourgeoisie develops over the long term, within an urban tradition stretching back to the Middle Ages. The emergence of a Luxembourgish metropolis is even more paradoxical considering the lack of urban tradition and the fact that industrial development was centred on one activity, the steel industry, located mostly in the southern part of the country. The economic history can be described in general in terms of a fundamental role being played by the state, but two distinct phases can be clearly discerned since the creation of the latter. Economic development has been associated with a territorial strategy and with the types of resources mobilised, which is reflected in the dynamics seen in a cross-border context – that of the Greater Region such as it exists at present (see Table C.1).

Table C.1 – Stages in Luxembourg's economic development

Period	Territorial strategy	Main economic activities	Economic process within the cross-border regional context	Resources mobilised	Urbanisation
1867-1970	Assertion	Industries	Similarity	Mineral resources	Double polarisation: Luxembourg and the mining region
1970-present	European integration	Financial and European administrative services	Differentiation and interface	Human resources	Metropolisation

Author: Reitel, 2010.

The independence of Luxembourg and its neutrality have been guaranteed since 1867, which was reflected in the dismantling of its fortifications. The state instituted "a clever financial policy" (Bauler 2001), designed to attract investors from neighbouring states that had begun their industrial development much earlier – not only the Germans and French, but also the Belgians. This stage in the emergence of an industrial economy was also the one that saw the territorial establishment of Luxembourg and the development of a national society. Luxembourg sought to become autonomous and distinguish itself from the neighbouring powerful states. This was done on the political level thanks to its neutrality, but on the economic level, development was based on the same factors as for the neighbouring states; the mining and steel industries subsequently became a factor of spatial continuity, which would come to the forefront with the creation of the ECSC. In the 1950s the urban network had a bipolar structure, with Luxembourg City concentrating the political functions and the southern region consisting of an urban network composed of small industrial centres. While the Luxembourgish state established its identity in political terms, in economic terms the territory resembled its neighbours due to its economic specialisation.

Metropolisation began from the 1960s onward and here, once again, the state played a major role. The state developed a niche policy within the EEC as a response to the crisis in the steel industry. This policy involved the adoption of "liberal" legislation encouraging the development of companies within the context of European integration. The aim was to maintain its independence in the European context by developing an economic strategy based on differentiation from its neighbours and exploitation of its position as an interface. The success of this policy exceeded all expectations. Thanks to its small size and its sovereignty, the Luxembourgish state was able to react rapidly within a changing context where openness is essential to attract capital and labour.

In economic terms, the development of Luxembourg now clearly distinguishes itself from that of its neighbours. In the current phase, one can clearly observe a phenomenon of concentration of activities on two levels. Growth in the job market in Luxembourg is significantly stronger than in the neighbouring countries, while within the country Luxembourg City is consolidating its leading role. The concentration of employment is reflected in ever-increasing economic and spatial polarisation across the neighbouring regions of Belgian Luxembourg, French Lorraine and German Saarland and Rhineland-Palatinate.

The aim of maintaining the autonomy of the nation has resulted in the emergence of a metropolis with unique characteristics. This territorial construction has become a metropolitan state: the development of metropolitan functions is not based on the existence of an urban bourgeoisie or the presence of companies, but rather it is the state that has implemented attractive conditions for economic activities, in particular financial ones. This has occurred within a context of European integration, which promotes free movement, thus facilitating the growth of mobility within the cross-border region. In this context, the border is no longer a barrier but rather a resource: Luxembourg has been able to exploit its differences and its position as an interface. The Luxembourgish state has only very recently become aware of this metropolitan dimension. The future prospects of this state-metropolis currently depend on the strategies implemented by the main actors on the basis of the issues as they perceive them.

The issues at stake in the construction of a metropolis: a development over time

The Luxembourgish metropolis is facing three key challenges for its future development. First, the metropolitan dynamic needs to be rendered sustainable. Second, the spatial development needs to be considered. Finally, the question about which mode of governance should be chosen for this spatio-temporal entity must be addressed.

Its small size and specialisation tends to give the Luxembourgish metropolis a certain fragility. Luxembourg has achieved recognition as a financial centre, but it has not escaped the threat of its activities being attracted to other, larger locations. This observation must, however, be qualified; the European institutions are not about to disappear and contribute significantly to the importance of the metropolis. However, at the same time the financial activities are linked to the attractiveness of the territory to businesses. From this perspective, the image of an "entrepreneurial city", which Luxembourg has been able to develop, is an undeniable advantage in terms of the competition between cities and between countries. The "entrepreneurial city" is a city conducive to business and the accumulation of capital, and which plays a role in driving wealth creation (Jessop and Sum, 2000). The desire of the state to have "liberal" legislation and an entrepreneurial climate have shaped the identity of Luxembourg. However, two questions remain unanswered. Does the metropolis deserve to grow in size? And should economic diversification be foreseen? The two questions are undoubtedly linked.

At stake is whether the increase in size will have more positive externalities than negative ones. The growth is only worthwhile if the metropolis passes a threshold that allows it to minimise certain aspects that contribute to its fragility. The risk of losing certain positive aspects – those concerning interpersonal relations – is a real one. The option is a viable one if the increase in employment in the financial sector results in improved economic performance and better visibility at the global level. An increase in population would also allow certain resources to be better exploited and urban dynamism to be reinforced. Demographic growth meets its full potential when reinforcing an urbanity dependent on both density and diversity (Lévy 1997).

The question of diversity also arises in economic terms. The fragility of Luxembourg is due not so much to its high degree of specialisation as to the weak presence of knowledge-intensive activities; even when they are present, these activities do not confer strong visibility on the European or global level. What is needed is the emergence of new activities favouring scientific innovation. However, until recently, Luxembourg has preferred to recruit highly qualified people educated in other countries. The creation of a university reflects a paradigm shift and a desire to create a labour force educated within the country and support endogenous development. While the importance of this has been clearly identified, its implementation remains fragile. The return on investment is not realised over a small number of years and the scientific progress cannot be achieved by decree. Developing links with the metropolis' economic fabric takes time.

Sustaining long-term prospects for the metropolis also requires increasing public awareness of the city. The image of Luxembourg has changed, but is its image really that of a metropolis rather than that of a microstate? The fact that the state and the capital city bear the same name leads to confusion, but this double usage of the name reinforces the idea of a state-metropolis. Furthermore, Luxembourg has undeniably become a cultural centre, but is it also considered to be a creative centre? The metropolis cannot content itself with being an economic centre or a centre of technical innovation; it is also a site of creativity where new ways of life emerge, where a certain lifestyle exists and where new forms of social conduct develop. In other words, it is a site where new forms of the urban are invented (Ascher 1995, Florida 2005). In contrast to major metropolises, Luxembourg does not appear to show signs of a creative urban milieu where new forms of art and culture appear (Grésillon 2002). However, creativity and innovation are two essential characteristics of a metropolis. One essential issue is to transform the urban centre into an urban space capable of attracting a creative population. This issue leads to the question of urban identity, which is the result of a combination of regional rootedness and openness to the world.

The second issue concerns the spatial extension of the metropolis. Metropolitan employment is largely concentrated in Luxembourg City, and, to a lesser extent, within the country. The neighbouring countries are primarily residential areas that provide certain services to the population of the metropolitan region. The complementarity of the functions of either side of the borders also reflects the existence of socio-economic disparities. These inequalities are not a feature specific to Luxembourg; they are characteristic of all the metropolises (Bassand 1997). The fact that each dyad also emphasises a socio-economic discontinuity is, however, a notable specific characteristic of the metropolis that the local authorities cannot ignore and which requires a multiscalar approach.

First, the heart of the metropolis requires specific attention. Urbanity should be reinforced within the districts where metropolitan employment is concentrated and which are characterised by a high degree of functional specialisation; this concerns primarily Kirchberg and to a lesser extent the city centre. The use of these areas outside of working hours, their capacity to generate urban vitality and to produce meaning deserve to be developed. The increase in residential density, the increase in the degree of residential diversity, and the development of dedicated public transport systems offer several ways in which the metropolitan centre can be strengthened and given a purpose that is based not only on its image as a centre of employment.

Secondly, the establishment of the University of Luxembourg on the Esch-Belval site will result in the emergence of a second centre of metropolitan employment at the national level. This centre will be one component of a larger metropolitan system. In order to maximise relations between the metropolitan actors and to create synergies, exchanges between the two centres will need to be maximised. The risk is that the institutions situated in Esch-Belval may feel that they are located within a periphery and not within a metropolitan system.

Finally, the cross-border dimension must not be ignored. The functional complementarities are the result of the existence of growing differences between employment and residence. At present, no authority has questioned this system. The growth of these differences may be, in the long term, prejudicial to the functioning of the metropolis, because it risks creating negative externalities, some of which can already be observed: increased property prices, traffic jams, and also risks of increased frustration and the emergence of symbolic social or cultural borders that cause tensions. Furthermore, several significant urban agglomerations can be reached in under an hour from Luxembourg and the political authorities that control these areas do not wish to confine themselves to a role as a service centre. Metz, Thionville, Saarbrücken, Trier and Arlon benefit from their proximity to Luxembourg but also experience certain negative impacts. Skilled workers are attracted to Luxembourg where salaries are higher, which undermines the urban economic bases. In addition, in the context of a growing metropolis, it can be asked what roles these agglomerations can play in order to enter into a process of reciprocal development (a win-win process). Ultimately the approach to the development of a metropolitan region is called sharply into question.

Questioning the territorial management of a cross-border urban region involves asking questions about the mode of governance. The Luxembourgish state is an essential actor within the organisation of the metropolitan region. However, the future development of the metropolis will require real synergies between the various actors within the metropolitan region, and in particular the local authorities of the neighbouring countries. Consequently, what is essential is the transformation of the state-metropolis into a true cross-border metropolis. This involves the development of a project corresponding to a shared vision of the future. However, the concept of the Cross-Border Polycentric Metropolitan Region proposed at the 11th summit of the Greater Region in 2009 by Luxembourg perhaps offers a new perspective. Beyond the question of the size of the area, which has already been raised (see Sohn and Walther, Chapter 11), the implementation of this ambition raises several questions, in relation to both the spatial interpretation and the organisation of the political power that it requires. Beyond the vision, it is the relationship between the Luxembourgish state and the other actors that is at stake at three different levels.

The first level is that of the agglomeration of Luxembourg, where the strengthening of the urbanity is one of the major requirements, calling for coordination between Luxembourg City, the state and the other municipalities within the urban area. The emergence of a political framework on this scale, with the City of Luxembourg playing a leadership role is perhaps necessary to take forward the development of this urban space. A second level relates to the national territory. The management of functional, economic, residential and recreational complementarities and their organisation is one of the major requirements. Ways must be found to distribute certain functions of the metropolis, while avoiding dispersion, which would be harmful to how it generally functions. Given the multiplicity of *communes*, one can imagine that these could be federated according to criteria that remain to be defined, in accordance with the aims of territorial development or considering a functional division. Finally, the third level concerns the cross-border space within which the future metropolis will develop. The contours of this area are at present hazy. This means that both a relevant project perimeter and the cross-border partnerships that will help to achieve a shared vision of development will need to be defined. The urban local authorities are likely to be key partners; this includes Luxembourg City, in order to define the future relationships, divisions of functions, etc.

The partnership of the regional and urban authorities within the cross-border polycentric metropolitan region (CBPMR) may in the long term result in the development of a system of regulation able to correct the distortions resulting from the varying dynamics on either side of the border. The Metroborder project financed by the European Spatial Planning Observation Network (ESPON), which will end in December 2010, should provide several ways to define in greater detail how the CBPMR could develop. Among the various terms used within the framework of the project, polycentrism is often used by the European institutions. This can be considered as a policy that aims to both increase coherence and recognise certain differences. However, this concept has given rise to various questions. Polycentrism is opposed to monocentrism (Reynaud 1981), the domination of one urban centre over another, but the concept can also be interpreted as the ability to coordinate the development of several centres while avoiding concentration within a single centre. Another issue relates to the political dimension that can be raised by polycentrism: the political powers involved in the development of a project and the interdependencies and complementarities that need to be established between the sites of political power. Polycentrism thus requires not only the existence of urban centres (Cattan 2007), but also the recognition of their ability to implement political projects and mobilise resources to achieve synergies. If clearly defined and accepted by the various actors, polycentrism can be one of the bases of a development policy for a cross-border area that is seeking to avoid too much concentration within the main centre. Such an option would not preclude its recognition on the functional level, nor the existence of possible leadership.

If it is the object of a concerted strategy, the development of a multiscalar and cross-border partnership, within which the Luxembourgish state provides leadership and monitors the relationships between actors, could provide an innovative model for governance within a European context. Thus, Luxembourg will no longer be seen simply as an economic model, but also as having a political project developed jointly with the neighbouring countries and that goes far beyond its national borders.

The Luxembourgish metropolis – a replicable model?

Ultimately, the way in which the Luxembourgish state has constructed a state-metropolis can be considered to be exemplary. The attractiveness of Luxembourg has been especially high over the last two decades within Europe and has allowed it to increase its visibility within the system of European cities. Two questions remain: first, can anything be learned from the Luxembourgish model? Second, is it an original spatial and temporal configuration?

Let us recall briefly several characteristics of this metropolis. First, this system was developed within a specific institutional and temporal context, that of the European project. As a buffer state, Luxembourg's position has been dependent since its creation on the relations between its large neighbours. The Franco-German reconciliation, and more generally the pacification of relations between states that had been in conflict, contributed to the opening up new prospects for development after the Second World War. The geo-political situation is one of the elements that contributed to providing the outlines for the metropolitan development. Luxembourg was able to exploit the advantages of its position as an interface on a cultural and linguistic border. Thanks to this dimension, Luxembourg succeeded in becoming known as a place where several languages were spoken fluently – French and German of course, but also, and above all, English. This ability to use multiple languages is certainly not the least important factor in the success of the Luxembourgish model. The free movement promoted by the process of European integration is another important element. In this context, the microstate has, while preserving its sovereignty and independence, succeeded in mobilising and attracting foreign labour and capital. Due to the small size of its market, Luxembourg has pursued specialisation within a European or global framework. The role of the state and the economic and political openness are reminiscent of Hong Kong and, above all, Singapore (Fau 1999, Breitung 2001). However, these countries have much larger populations and a greater diversity of activities.

Can this model be observed within other regions of the world? The prima facie response would seem to be negative. Europe is the only region of the world to have experienced such a great degree of regional integration. In addition, it would be necessary to identify microstates that are also buffer states and with a possible cross-border dimension. From this perspective, Singapore has the greatest resemblance to the Luxembourgish model, due to its position as an interface, the proximity of two larger countries with greater populations, and certain cross-border aspects (Fau *ibid.*). The model appears to be transferable only when there is a state power with the ability to legislate in a context of peaceful and open borders. In this context, it is within the federal states that small, densely populated and urbanised regions can be identified, which are heavily integrated globally and are characterised by the existence of entrepreneurialism, metropolitan dynamics and cross-border integration. It is thus in Switzerland that models closest to those of Luxembourg can be found. In terms of the three characteristics cited, Basel and Geneva present undeniable convergences with Luxembourg, despite the differences in their urban histories and spatial forms. With these possible exceptions, and with several nuances, the Luxembourgish model appears to be relatively unique.

References

Agnew J. 2000. From the political economy of regions to regional political economy. *Progress in Human Geography*, 24: 101-110.

Akademie für Raumforschung und Landesplanung 2007. *Metropolregionen. Innovation, Wettbewerb, Hundlungsfähigkeit*. ARL Positionspapier, 71.

Albrechts L. 2006a. Bridge the gap: from spatial planning to strategic projects. *European Planning Studies*, 14(10): 1487-1500.

Albrechts L. 2006b. Shifts in Strategic Planning? Some evidence from Europe and Australia. *Environment and Planning A*, 38: 1149-1170.

Allain R., Baudelle G., Guy C. 2003. *Le polycentrisme, un projet pour l'Europe*. Rennes: Presses universitaires de Rennes.

Allen J., Cochrane A. 2007. Beyond the territorial fix: regional assemblages, politics and power. *Regional Studies*, 41(9): 1161-1175.

Amin A., Thrift N. 1995. Globalisation, institutional "thickness" and the local economy. In Healey P., Cameron S., Davoudi S., Graham S., Madani-Pour A. (eds.) *Managing cities. The new urban context*. Chichester: John Wiley, 91-108.

Anderson M. 1996. *Frontiers: territory and state formation in the modern world*. Cambridge: Cambridge Polity Press.

Anderson J., O'Dowd L. 1999. Borders, border regions and territoriality: contradictory meanings, changing significance. *Regional Studies*, 33(7): 593-604.

Anderson J., O'Dowd L., Wilson T.M. (eds.) 2003. *New borders for a changing Europe: cross-border co-operation and governance*. London: Routledge.

Anderson A., Wever E. 2003. Borders, border regions and economic integration: one world, ready or not. *Journal of Borderlands Studies*, 18(1): 27-38.

Anselin L. 1995. Local indicators of spatial association – LISA. *Geographical Analysis*, 27: 93-115.

Anselin L. 1996. The Moran scatterplot as an ESDA tool to assess local instability in spatial association. In Fischer M., Scholten H.J., Unwin D. (eds.) *Spatial analytical perspectives on GIS*. London: Taylor & Francis, 111-125.

Anselin L., Syabri I., Kho Y. 2006. GeoDa: an introduction to spatial data analysis. *Geographical Analysis*, 38: 5-22.

Antier G. 2005. *Les stratégies des grandes métropoles: enjeux, pouvoirs et aménagement*. Paris: Armand Colin.

Armstrong H., Read R. 1998. Trade and growth in small states: the impact of global trade liberalisation. *World Economy*, 27(4): 563-585.

Arthur W.B. 1994. *Increasing returns and path dependence in the economy*. Ann Arbor: University of Michigan Press.

Ascher F. 1995. *Métapolis ou l'avenir des villes*. Paris: Odile Jacob.

Ascher F. 2001. *Les nouveaux principes de l'urbanisme*. La Tour d'Aigues: Aube.

Ascher F. 2003. Métropolisation. In Lévy J., Lussault M. (eds.) *Dictionnaire de la géographie et de l'espace des sociétés*. Paris: Belin, 612-615.

Auburtin E. 2002. *Dynamiques et représentations transfrontalières de la Lorraine. Analyse géopolitique régionale appliquée*. Unpublished PhD dissertation, University of Paris VIII.

Bäck H., Heinelt H., Magnier A. (eds.) 2006. *The European mayor: political leaders in the changing context of local democracy*. Wiesbaden: VS Verlag.

Bade K.J. 2003. *Migration in European history*. London: Blackwell.

Bae C.H.C. 2003. Tijuana-San Diego: globalization and the transborder metropolis. *The Annals of Regional Science*, 37: 463-477.

Baert T. 2008. A cross-border approach to city-region (metropolitan) policies: the case of Lille Metropole. *Urban Research and Practice*, 1(2): 199-209.

Bailly A. 1987. Une géopolitique de la Regio Genevensis. *Geographica Helvetica*, 3: 191-202.

Bailly A. 1999. Regards croisés sur un concept. In Lacour C., Puissant S. (eds.) *La métropolisation. Croissance, diversité, fractures*. Paris: Anthropos, 153-164.

Bairoch P. 1996. *De Jericho à Mexico. Villes et économie dans l'histoire*. Paris: Gallimard.

Bar-El R., Parr J.B. 2003. From metropolis to metropolis-based region: the case of Tel-Aviv. *Urban Studies*, 40(1): 113-125.

Bassand M. 1997. *Métropolisation et inégalités sociales*. Lausanne: Presses polytechniques et universitaires romandes.

Bassand M., Kaufmann V., Joye D. (eds.) 2001. *Enjeux de la sociologie urbaine*. Lausanne: Presses polytechniques et universitaires romandes.

Bathelt H., Malmberg A., Maskell P. 2004. Clusters and knowledge: local buzz, global pipelines and the process of knowledge creation. *Progress in Human Geography*, 28: 31-56.

Bauer G., Roux J.-M. 1976. *La rurbanisation ou la ville éparpillée*. Paris: Seuil.

Bauler A. 2001. *Les fruits de la souveraineté nationale. Essai sur le développement de l'économie luxembourgeoise de 1815 à 1999. Une vue institutionnelle*. Luxembourg: Caisse centrale Raiffeisen.

Baumont C., Ertur C., Le Gallo J. 2004. Spatial analysis of employment and population density: the case of the agglomeration of Dijon, 1999. *Geographical Analysis*, 36(2): 146-176.

Bell D., Jayne M. 2006. *Small cities. Urban experience beyond the metropolis*. London: Routledge.

Benhamou S. 2004. Fragmentation des territoires et dynamiques de ségrégation urbaine. In Fitoussi J.-P., Laurent E., Maurice J. (eds.) *Ségrégation urbaine et intégration social*. Paris: La Documentation française, 295-312.

Berger F. 2005. Développement de l'emploi transfrontalier et portrait sociodémographique des frontaliers. *Population & Emploi*, 8, CEPS/INSTEAD.

Berry B.J.L. 1976. *Urbanisation and counter-urbanisation*. London: Sage.

Berry B.J.L., Cohen P.G., Goldstein H. 1968. Metropolitan area definition: a re-evaluation of concept and statistical practice. *Working Paper*, 28. Washington, DC: Bureau of the Census.

Bertin J. 1967. *Sémiologie graphique*. La Haye, Paris: Mouton, Gauthier-Villars.

Blatter J. 1996. *Political co-operation in cross-border regions: two explanatory approaches*. Paper presented at the European Regional Science Association 36th European Congress, ETH Zurich.

Blatter J. 2003. Beyond hierarchies and networks: institutional logics and change in transboundary spaces. *Governance*, 16(4): 503-526.

BMVBS 2011. *Metropolitane Grenzregionen*. Abschlussbericht des Modellvorhabens der Raumordung (MORO) "Überregionale Partnershaften in grenzüberschreitenden Verflechtsräumen". Bonn: Bundesamt für Bauwesen und Raumordnung.

Boiteux-Orain C., Guillain R. 2004. Changes in the intrametropolitan location of producer services in Île-de-France (1978-1997): Do information technologies promote a more dispersed spatial pattern? *Urban Geography*, 25(6): 550-578.

Bolt G., van Kempen R., van Ham M. (2008). Minority ethnic groups in the Dutch housing market: spatial segregation, relocation dynamics and housing policy. *Urban Studies*, 45(7): 1359-1384.

Bontje M. 2001. Dealing with deconcentration: population deconcentration and planning response in polynucleated urban regions in North-West Europe. *Urban Studies*, 38(4): 769-785.

Bonvalet C. 2005. Logement et vie familiale, un parcours résidentiel en mutations. *Informations sociales*, 3(123): 56-65.

Bonvalet C., Fribourg A.-M. 1990. *Stratégies résidentielles*. Paris: INED-Plan Construction et Architecture-MELTM.

Boschma R.A. 2005. Proximity and innovation: a critical assessment. *Regional Studies*, 39(1): 61-74.

Bottai M., Barsotti O. 2003. Les déplacements quotidiens: approches et modèles. In Caselli G., Vallin J., Wunsch G. (eds.) *Démographie: analyse et synthèse. Les déterminants de la migration*. Paris: INED, 173-197.

Bouquet T., Ousey B. 2008. *Cold steel: Britain's richest man and the multi-billion-dollar battle for a global industry*. London: Little Brown.

Bourdin A. 2005. *La métropole des individus*. Paris: Éditions de l'Aube.

Bourgain A., Pieretti P. 2006. Measuring agglomeration forces in a financial center. *Economics Bulletin*, 18(3): 1-9.

Braudel F. 1979. *Civilisation matérielle, économie et capitalisme, XVe-XVIIIe siècles. Tome 1. Les structures du quotidien: le possible et l'impossible*. Paris: Armand Colin.

Breitung W. 2001. *Hongkong und der Integrationsprozess. Räumliche Strukturen und planerische Konzepte*. Basel: Wepf & Co.

Brenner N. 2003. Metropolitan reform and the rescaling of state space in contemporary Western Europe. *European Urban and Regional Studies*, 10(4): 297-324.

Brenner N. 2004. Urban governance and the production of new state spaces in Western Europe, 1960-2000. *Review of International Political Economy*, 11(3): 447-488.

Brenner N., Theodore N. 2002. Cities and the Geographies of "Actually Existing Neoliberalism". *Antipode*, 34(3): 349-379.

Brinkley I., Lee N. 2008. *The knowledge economy in Europe*. London: The Work Foundation.

Bronner A.-C., 2009. L'accessibilité européenne. In Bousch P., Chilla T., Gerber P., Klein O., Schulz C., Sohn C., Wiktorin D. (eds.) *Atlas du Luxembourg*. Cologne: Emons, 134-135.

Brown L.A., Holmes J. 1971. The delimitation of functional regions, nodal regions, and hierarchies by functional distance approaches. *Journal of Regional Science*, 11: 57-72.

Brun J., Bonvalet C. 2002. Approches quantitatives ou qualitatives de la mobilité résidentielle: éléments de bilan et perspectives. *Espace, Populations, Sociétés*, 1-2: 63-72.

Brunet R. 1989. *Les villes européennes*. Paris: La Documentation française.

Brunet R., Ferras R., Théry H. 1993. *Les mots de la géographie. Dictionnaire critique*. Montpellier: Reclus – La Documentation française.

Brunet-Jailly E. 2000. Globalization, integration, and cross-border relations in the metropolitan area of Detroit (USA) and Windsor (Canada). *International Journal of Economic Development*, 2(3): 379-401.

Budd L. 1999. Globalisation and the crisis of territorial embeddedness. In Martin M. (ed.) *Money and the space economy*. Chichester: John Wiley, 115-138.

Burgers J., van der Lugt H. 2006. Spatial assimilation of minority groups: The case of suburbanizing Surinamese in the Rotterdam region. *Journal of Housing and Built Environment*, 21: 127-139.

Burnotte-Boreux C., Casteels Y. 2003. *La société luxembourgeoise à travers le recensement de 2001*. Luxembourg: STATEC.

Buursink J. 2001. The binational reality of border-crossing cities. *GeoJournal*, 54(1): 7-19.

Carillo F. (ed.) 2006. *Knowledge cities: approaches, experience and perspectives*. New York: Butterworth-Heinemann.

Carpentier S. 2007. *Mobilité quotidienne et ancrage résidentiel. Différenciation des pratiques spatiales et des représentations sociales selon la structure urbaine. L'exemple du Luxembourg*. Unpublished PhD thesis, University Louis Pasteur, Strasbourg.

Carpentier S. (ed.) 2010. *La mobilité résidentielle transfrontalière entre le Luxembourg et ses régions voisines*. Luxembourg: St-Paul.

Carpentier S., Gerber P. 2007. Les déplacements domicile-travail: en voiture, en train ou à pied? *Vivre au Luxembourg*, 3(53), CEPS/INSTEAD.

Carpentier S., Gerber P. 2009. De la mobilité résidentielle à la construction des espaces de la vie quotidienne. Une première approche. *Recherche Transports Sécurité*, 25(102): 61-72.

Cassis Y. 2008. *Les capitales du capital. Histoire des places financières internationales 1780-2005*. Paris: Champion.

Castells M. 1989. *The informational city: information, technology, economic restructuring and the urban-regional process*. Oxford: Blackwell.

Castells M. 1999. Grassrooting the space of flows. *Urban Geography*, 20(4): 294-302.

Castells M. 2003. The process of urban social change. In Cuthbert A.R. (ed.) *Designing cities. Critical readings in urban design*. Oxford: Blackwell, 23-27.

Cattan N. (ed.) 2007. *Cities and networks in Europe. A critical approach of polycentrism*. Montrouge: John Libbey Eurotext.

Cauvin C., Escobar F., Serradj A. 2007. *Cartographie thématique 2. Des transformations incontournables*. Paris: Hermès.

Cauvin C., Reymond H. 1991. Interaction spatiale et cartographie: les solutions de W. Tobler. *Espace, Populations, Sociétés*, 3: 467-485.

CEPS/INSTEAD 2008. *Suivi du développement territorial du Luxembourg à la lumière des objectifs de l'IVL*. Report for the Ministry of Spatial Planning, Luxembourg.

CES Lorraine 2007. *L'enjeu transfrontalier au cœur du développement de la Lorraine*. Plenary Session June 26th. Conseil économique et social de Lorraine. http://www.ces.public.lu/fr/collaborations/rapp-ces-lorraine.pdf.

CES Lorraine 2009. *Situation économique, sociale et environnementale de la Lorraine 2008-2009*. Plenary Session 24th November. http://ces.lorraine.eu/jahia/webdav/shared/Rapports/2009/Situation Economique Sociale et Environ nementale de la Lorraine 2008-2009 2024.11.2009.pdf.

Chen X. 2005. *As borders bend. Transnational spaces on the Pacific Rim*. Lanham: Rowman & Littlefield.

Christaller W. 1933. *Die zentralen Orte in Süddeutschland*. Jena: Gustav Fischer.

City of London 2008. *The global financial centres index*. London: City of London.

Clark G.L., O'Connor K. 1997. The informational content of financial products and the spatial structure of the global finance industry. In Cox KR. (ed.) *Spaces of globalization: reasserting the power of the local*. New York: Guildford, 89-114.

Clark G.L, Wójcik D. 2005. Path dependence and financial markets: the economic geography of the German model, 1997-2003. *Environment and Planning A*, 37: 1769-1791.

Clement N.C. 1997. The changing economics of international borders and border regions. In Ganster P., Sweedler A., Scott J., Dieter-Eberwein W. (eds.) *Borders and border regions in Europe and North America*. San Diego: San Diego State University Press and Institue for Regional Studies of the Californias, 47-63.

Clément F. 2008. La construction sociale du territoire de la Grande Région: une confusion entre les concepts de collaboration et d'intégration. *Gouvernance & Emploi*, 2, CEPS/INSTEAD.

Coffey W. 2000. The geographies of producer services. *Urban Geography*, 21(2): 170-183.

Cook A.K. 1991. Metropolitan dominance versus decentralisation in the information age. *Social Science Quarterly*, 72(2): 284-298.

Coombes M.G., Wymer C. 2001. A new approach to identifying localities: representing "places" in Britain. In Madanipour A., Hull A., Healey P. (eds.) *The governance of place: space and planning processes*. Aldershot: Ashgate, 51-69.

Courlet C. 1988. La frontière: coupure ou couture? *Économie et Humanisme*, 301: 5-12.

Da Cunha A., Both J.-F. 2004. *Métropolisation, villes et agglomérations. Structures et dynamiques socio-démographiques des espaces urbains*. Neuchâtel: Swiss Federal Statistical Office.

Dabinett G., Richardson T. 2005. The europeanization of spatial strategy: shaping regions and spatial justice through governmental ideas. *International Planning Studies*, 10(3-4): 201-218.

Dallenne P. 2009. Les mutations économiques et spatiales en Europe depuis 1945. In Dallenne P., Nonjon A. (eds.) *L'espace mondial, fractures ou inter-dépendances?* Paris: Ellipses.

DATAR 2004. *L'offre métropolitaine française vue par les emplois métropolitains supérieurs*. Paris: DATAR, Fédération nationale des agences d'urbanisme.

Davoudi S. 2003. Polycentricity in European spatial planning: from an analytical tool to a normative agenda. *European Planning Studies*, 11(8): 979-999.

Davoudi S. 2008. Conceptions of the city-region: a critical review. *Proceedings of the ICE – Urban Design and Planning*, 161(2): 51-60.

De Boe P., Grasland C., Healy A. 1999. *Spatial integration*. Study Programme on European Spatial Planning, Final report 1.4.

Decoville A. 2008, Métropolisation et résistance des territoires: l'exemple luxembourgeois. *L'Espace politique*, 4: 53-68.

Del Biondo L. 2009. La conversion des friches sidérurgiques à la frontière franco-luxembourgeoise: un enjeu transfrontalier. *L'Information géographique*, 73(3): 46-64.

Deloitte 2006. *Etude d'impact de l'industrie financière sur l'économie luxembourgeoise*. Deloitte Luxembourg Report, Luxembourg.

Dematteis G. 2003. Centralité. In Lévy J., Lussault M. (eds.) *Dictionnaire de la géographie et de l'espace des sociétés*. Paris: Belin, 139-141.

Denys C. 2002. Construction et image de la frontière entre les Pays-Bas et la France: une réflexion historique sur la relation entre la frontière et la ville frontalière. In Reitel B., Zander P., Piermay J.-L., Renard J.-P. (eds.) *Villes et frontières*. Paris: Anthropos, 179-187.

Dessemontet P., Perlik M., Schuler M. 2005. Metropolitanregionen versus Kantone. In Blöchliger H. (ed.) *Baustelleföderalismus*. Zurich: Avenir Suisse and NZZ, 156-165.

Dezert B., Metton A., Steinberg J. 1991. *La périurbanisation en France*. Paris: Sedes.

Dormois R. 2008. Les coalitions dans l'analyse des politiques urbaines post-keynésiennes. *Métropoles*, 4. http://metropoles.revues.org/document 3122.html.

DTZ 2007. *Global office occupancy cost survey*. London: DTZ.

Dupuy G. 1991. *L'urbanisme des réseaux: théories et méthodes*. Paris: Armand Colin.

Dupuy G. 1999. *La dépendance automobile: symptômes, analyses, diagnostic, traitements*. Paris: Anthropos.

Duranton G., Puga D. 2005. From sectoral to functional urban specialisation. *Journal of Urban Economics*, 57(2): 343-370.

e-BIRD 2005. *Comprendre les bassins de main-d'œuvre de la Grande Région.* Conference proceedings "Détermination des bassins de main-d'œuvre fonctionnels", Luxembourg 14th November.

Enaux C., Gerber P. 2008. Les déterminants de la représentation transnationale du bassin de vie. Une approche fondée sur l'attachement au lieu des frontaliers luxembourgeois. *Revue d'économie régionale et urbaine*, 5: 725-752.

ESPON 2005. *Potentials for polycentric development in Europe*, 1.1.1. Luxembourg: ESPON.

ESPON 2006. *The modifiable areas unit problem, 3.4.3*. Final Report. Luxembourg: ESPON.

ESPON 2007. *Study on urban functions*, 1.4.3. Final Report. Luxembourg: ESPON.

ESPON 2010. *Metroborder. Cross-border polycentric metropolitan regions.* Interim Report. Luxembourg: ESPON.

European Commission 1999. *European spatial development perspective. Towards balanced and sustainable development of the territory of the European Union.* Luxembourg: Office for Official Publications of the European Communities.

European Union 2007a. *Leipzig charter on sustainable European cities*. European Union.

European Union 2007b. *Territorial agenda of the EU.* European Union.

Eurostat 2006. *High-technology manufacturing and knowledge-intensive services sectors*. Eurostat Metadata. http://europa.eu/estatref/info/sdds/en/htec/htec_agg_nace.pdf.

Fau N. 1999. Hong-Kong et Singapour, des métropoles transfrontalières. *L'Espace géographique*, 3: 241-255.

Faulconbridge J. 2004. London and Frankfurt in Europe's evolving financial centre network. *Area*, 36(3): 235-244.

Faulconbridge J., Engelen E., Hoyler M., Beaverstock J. 2007. Analysing the changing landscape of European financial centres: the role of financial products and the case of Amsterdam. *Growth and Change*, 38(2): 279-303.

Fischman R. 1998. Beyond utopias: urbanism after the end of cities. *Urbanitats*, 7. http://www.publicspace.org/en/text-library/eng/a004-beyond-utopia-urbanism-after-the-end-of-cities.

Florida R. 2005. *Cities and the creative class*. New York and London: Routledge.

Foucher M. 1991. *Fronts et frontières. Un tour du monde géopolitique*. Paris: Fayard.

Fourastié J. 1989. *Le grand espoir du XXe siècle*. Paris: Gallimard.

Franz N. 2005. Der Finanzplatz Luxemburg als Ergebnis wirtschaftlichen Bedarfs, politische Willens und europäischer Integration. In Merki C.M. (ed.)

Europas Finanzzentren. Geschichte und Bedeutung im 20. Jahrhundert. Frankfurt am Main: Campus, 149-165.

Friedmann J. 1986. The world city hypothesis. *Development and Change*, 17: 69-83.

Fujita M., Thisse J.-F. 2002. *Economics of agglomeration. Cities, industrial location and regional growth.* Cambridge: Cambridge University Press.

Genestier P., Jouve B., Boino P. 2008. L'aménagement urbain et territorial, entre post-keynésianisme et néo-keynésianisme. *Métropoles*, 4. http://metropoles. revues.org /document3583.html.

Gerber P. 2000. *Confort postmoderne et gentrification, éléments émergents de nouvelles centralités. Le cas de Strasbourg.* Unpublished PhD thesis, University Louis Pasteur, Strasbourg.

Gerber P., Carpentier S., Petit S., Pigeron-Piroth I. 2008. Mobilités quotidienne et résidentielle au Luxembourg: un aperçu à travers l'outil MobilluxWeb. *Population & Territoire*, 13, CEPS/INSTEAD.

Gerber P., Klein O. 2009. Un phénomène de "city" marqué à Luxembourg-Ville. In Bousch P., Chilla T., Gerber P., Klein O., Schulz C., Sohn C., Wiktorin D. (eds.) *Atlas du Luxembourg*, Cologne: Emons, 62-63.

Giddens A. 1981. *A contemporary critique of historical materialism. Vol. 1: Power, property and the state.* London: Macmillan.

Gildersleeve C.R. 1978. *The international border city: urban spatial organization in a context of two cultures along the United States – Mexico boundary.* Unpublished PhD thesis, University of Nebraska.

Glanzman L., Grillon N., Kruse C., Thierstein A. 2004. Polycentricity and metropolitan governance. A Swiss case study. *ERSA Conference Paper* 440. http://www-sre.wu-wien.ac.at/ersa/ersaconfs/ersa04/PDF/440.pdf.

Gløersen E., Lähteenmäki-Smith K., Dubois A. 2007. Polycentricity in transnational planning initiatives: ESDP applied or ESDP reinvented? *Planning, Practice and Research*, 22 (3): 417-437.

Gordon I.R., McCann P. 2000. Industrial clusters: complexes, agglomerations and/or social networks? *Urban Studies*, 37(3): 513-532.

Greater Region 2009. *Déclaration commune.* Luxembourg: Presidence of the 11th Summit of the Greater Region, 17th July.

Greenfield H.I. 1966. *Manpower and the growth of producer services.* New York: Columbia University Press.

Grésillon B. 2002. *Berlin, métropole culturelle.* Paris: Belin.

Grix J., Knowles V. 2003. The Euroregion and the maximization of social capital: Pro Europa Viadrina. In Anderson J., O'Dowd L., Wilson T.M. (eds.) *New borders for a changing Europe. Cross-border cooperation and governance.* London: Frank Cass, 154-176.

Grote M. 2007. Mobile marketplaces – Consequences of the changing governance of European stock exchange. *Growth and Change*, 38(2): 260-278.

Grote M. 2008. Foreign bank's attraction to the financial centre Frankfurt – an inverted "U"-shaped relationship. *Journal of Economic Geography*, 8(2): 239-258.

Grote M. 2009. Financial centers between centralization and virtualization. In Alessandrini P., Fratianni M., Zazzaro A. (eds.) *The changing geography of banking and finance*. New York: Springer, 277-294.

Grote M., Lo V., Harrschar-Ehrnborg S. 2002. A value chain approach to financial centres. The case of Frankfurt. *Tijdschrift voor Economische en Sociale Geografie*, 93(4): 412-423.

Guillain R., Le Gallo J., Boiteux-Orain C. 2006. Changes in spatial and sectoral patterns of employment in Île-de-France, 1978-97. *Urban Studies*, 43(11): 2075-2098.

Häkli J. 2009. Boundaries of trust: building a transnational space in Haparanda-Tornio. In Häkli J., Minca C. (eds.) *Social capital and urban networks of trust*. Farnham: Ashgate, 205-232.

Halbert L. 2004. The intrametropolitan decentralization of business services in the Paris Region: patterns, interpretation, consequences. *Economic Geography*, 80(4): 381-405.

Halbert L., Pain K., Thierstein A. 2006. European polycentricity and emerging mega-city regions: "one size fits all" policy? *Built Environment*, 32(2): 206-218.

Hall P., Pain K. (eds.) 2006. *The polycentric metropolis. Learning from mega-city regions in Europe*. London: Earthscan.

Hamnett C. 2003. Gentrification and the middle-class remaking of Inner London, 1961-2000. *Urban Studies*, 40(12): 2401-2426.

Hampton M.K. 1996. Sixties child? The emergence of Jersey as an offshore finance centre 1955-71. *Accounting, Business and Financial History*, 6(1): 51-71.

Hansen N. 1977. Border regions: a critique of spatial theory and a European case study. *The Annals of Regional Science*, 11(1): 1-14.

Hansen N. 1981. *The border economy*. Austin: University of Texas Press.

Hansen P.A, Serin G. 2007. Integration strategies and barriers to co-operation in cross-border regions: case study of the Øresund Region. *Journal of Borderlands Studies*, 22(2): 39-56.

Harding A., Marvin S., Robson B. 2006. *A framework for city-regions*. London: Office of the Deputy Prime Minister.

Hatton T.J., Tani M. 2005. Immigration and inter-regional mobility in the UK, 1982-2000. *Economic Journal*, 115(507): 342-358.

Hatzichronoglou T. 1997. Revision of the high-technology sector and product classification. *OECD Science, Technology and Industry Working Papers*, 2. Paris: OECD.

Haumont N. (ed.) 2000. *La ville, agrégation et ségrégation sociales*. Paris: L'Harmattan.

Helweg-Nottrot I. 2001. *Kirchberg 1961-2001*. Luxembourg: Fonds d'urbanisation et d'aménagement du Plateau du Kirchberg.

Hermelin B. 2007. The urbanization and suburbanization of the service economy: producer services and specialization in Stockholm. *Geografiska Annaler: Series B, Human Geography*, 89: 59-74.

Herzog L.A. 1990. *Where North meets South. Cities, space and politics on the U.S.-Mexico border*. Austin: Center for Mexican American Studies, University of Texas.

Herzog L.A. 1991. Cross-national urban structure in the era of global cities: the US-Mexican transfrontier metropolis. *Urban Studies*, 28(4): 519-533.

Heywood I. 1998. *Introduction to Geographical Information Systems*. New York: Addison Wesley Longman.

Hodgson G.M. 2006. What are institutions? *Journal of Economic Issues*, 40(1): 1-25.

Hooghe L. (ed.) 1996. *Cohesion policy and European integration: building multi-level governance*. Oxford: Oxford University Press.

Horner M.W., Murray A.T. 2002. Excess commuting and the modifiable areal unit problem. *Urban Studies*, 39(1): 131-139.

Howard E. 1985. *Garden cities of to-morrow*. London: Eastbourne.

Hübsch M. 2004. Wirtschaftsräumliche Möglichkeiten und Grenzen einer wirtschaftsfördernden Politik für das Großherzogtum Luxemburg. *Aachener Geographische Arbeiten*, 39.

Husson C. 2002. *L'Europe sans territoire*. La Tour d'Aigues, Paris: Aube, DATAR.

Innenministerium 2004. *Ein Integratives Verkehrs-und Landesentwicklungskonzept für Luxembourg*. Luxembourg: DATUR.

INSEE 2003. *Structuration de l'espace rural: une approche par les bassins de vie*. Report for the DATAR. http://www.insee.fr/fr/ffc/docs_ffc/bassins_vie/rapport.htm.

INSEE 2009a. Les espaces urbains lorrains: entre agglomération et dispersion. *INSEE Économie*, 121-122.

INSEE 2009b. Les revenus des ménages du nord lorrain: "boostés" par les salaires frontaliers, *Économie lorraine*, 154.

INSEE, 2010. Metz-Thionville: une aire métropolitaine à construire. *Économie lorraine*, 201.

INSEE-OCSTAT 2008. *Synthèse 2008*. Observatoire statistique transfrontalier des accords bilatéraux. Geneva, Lyon: OCSTAT, INSEE.

Institut für Ländliche Strukturforschung (IfLS) 2001. *Schéma de développement de l'espace Saar-Lor-Lux +*. Frankfurt am Main: IfLS.

Jackson K.T. 1985. *Crabgrass frontier: the suburbanization of the United States*. New York: Oxford University Press.

Jäger S. 1999. *Kritische Diskursanalyse. Eine Einführung*. Duisburg: DISS.

Jasso M. 2007. *Competition and cooperation of the cities: Vienna-Bratislava metropolitan region*. Paper presented at the 2nd Central European Conference in Regional Science, Nový Smokovec, 10-13th October.

Jessop B. 2002. Liberalism, neoliberalism and urban governance: a state-theoretical perspective. *Antipode*, 34: 452-472.

Jessop B. 2004. The European Union and recent transformations in statehood. In Riekmann S.P., Mokre M., Latzer M. (eds.) *The state of Europe: transformations of statehood from a European perspective*, Frankfurt am Main: Campus, 75-94.

Jessop B., Sum N.-L. 2000. An entrepreneurial city in action: Hong Kong's emerging strategies in and for (inter) urban competition. *Urban Studies*, 37(12): 2287-2313.

Jouve B. 1996. L'agglomération franco-genevoise: entre crise de voisinage et reconfiguration territoriale. *L'Espace géographique*, 1: 65-76.

Julien P., 1995a. La "métropolarisation" des actifs structure le territoire. *Économie et Statistique*, 290: 33-49.

Julien P. 1995b. French towns, higher urban functions and strategic employment. *Urban Studies*, 32(2): 401-423.

Julien P. 2002. Onze fonctions pour qualifier les grandes villes. *INSEE Première*, 840.

Karlson C., Olsson M. 2006. The identification of functional regions: theory, methods, and application. *Annals of Regional Science*, 40: 1-18.

Karreman B. 2009. Financial geographies and emerging markets in Europe. *Tijdschrift voor Economische en Sociale Geografie*, 100(2): 260-266.

Kaufmann V. 2000. *Mobilité quotidienne et dynamiques urbaines. La question du report modal*. Lausanne: Presses polytechniques et universitaires romandes.

Kaufmann V. 2005. Mobilités et réversibilités: vers des sociétés plus fluides? *Cahiers internationaux de sociologie*, 118: 119-135.

Kaufmann V. 2008. *Les paradoxes de la mobilité: bouger, s'enraciner*. Lausanne: Presses polytechniques et universitaires romandes.

Kearns A., Paddison R. 2000. New challenges for urban governance. *Urban Studies*, 37(5): 845-850.

Keating M., Loughlin J. (eds.) 1997. *The political economy of regionalism*. London: Frank Cass.

Keller R., Hirseland A., Schneider W., Viehöver W. 2001. *Handbuch Sozialwissenschaftliche Diskursanalyse. Band 1: Theorien und Methoden*. Opladen: Leske and Budrich.

Kirsch R. 1994. Naissance et essor d'une place financière. In Trausch G. (ed.) *La Ville de Luxembourg. Du château des comtes à la métropole européenne*. Antwerpen: Mercator, 285-295.

Klagge B. 2009. Finanzmärkte, Unternehmensfinanzierung und die aktuelle Finanzkrise. *Zeitschrift für Wirtschaftsgeographie*, 53(1-2): 1-13.

Klein O. 2007. *Modélisation et représentations spatio-temporelles des déplacements quotidiens urbains. Application à l'Aire Urbaine Belfort-Montbéliard*. Unpublished PhD thesis, University Louis Pasteur, Strasbourg.

Kloosterman R.C., Musterd S. 2001. The polycentric urban region: towards a research agenda. *Urban Studies*, 38(4): 623-633.

Knieling J. (ed.) 2009. *Metropolregionen. Innovation, Wettbewerb, Handlungsfähigkeit, Forschung und Sitzungsberichte der ARL*. Band 231. Hannover: Akademie für Raumforschung und Landesplanung.

Korcelli-Olejniczak E. 2007. Metropolitan functions as a research question: definitions and identifications in the ESPON and alternative approches. GaWC Research Bulletin 248, http://lboro.ac.uk/gawc/rb/rb248.htlm.

König W., Schamp E.W., Beck R., Handke M., Prifling M., Späthe S., Vykoukal J. 2007. *Finanzcluster Frankfurt. Eine Clusteranalyse am Finanzzentrum Frankfurt/Rhein-Main*. Norderstedt: Books on Demand.

Krätke S. 2007. Metropolisation of the European economic territory as a consequence of increasing specialisation of urban agglomeration in the knowledge economy. *European Planning Studies*, 15(1): 1-27.

Krugman P. 1991. Increasing returns and economic geography. *Journal of Political Economy*, 99(3): 483-499.

Krugman P. 1998. What's new about the new economic geography? *Oxford Review of Economic Policy*, 14(2): 7-17.

Kruse C. 2005. *Börsengänge am Finanzplatz Schweiz. Vernetzte Finanzintermediäre als Erfolgs-oder Risikofaktor für Börsenunternehmen?* Unpublished PhD thesis, ETH Zurich.

Lacoste Y. 2003. *De la géopolitique aux paysages. Dictionnaire de géographie*. Paris: Armand Colin.

Lacour C., Puissant S. 1999. *La métropolisation. Croissance, diversité, fractures*. Paris: Anthropos.

Laganier J., Vienne D. 2009. Recensement de la population de 2006. *INSEE Première*, 1218.

Laulajainen R. 1998. *Financial geography. A banker's view*. London: Routledge.

Le Breton E. 2005. *Bouger pour s'en sortir. Mobilité quotidienne et intégration sociale*. Paris: Armand Colin.

Le Galès P. 2003. *Le retour des villes européennes. Sociétés urbaines, mondialisation, gouvernement et gouvernance*. Paris: Presses de Sciences Po.

Le Galès P. 2004. A chaque ville sa gouvernance. La transformation des politiques urbaines. *Sciences Humaines*, 44: 34-39.

Le Galès P., Harding A. 1998. Cities and states in Europe. *West European Politics*, 21(3): 120-145.

Lee R. 2009. Integration. In Gregory D., Johnston R., Pratt G., Watts M.J., Whatmore S. (eds.) *The dictionnary of human geography*. Chichester: Wiley-Blackwell, 387-388.

Lefèvre C. 2004. *Les coopérations métropolitaines en Europe*. Paris: DATAR.

Legrand M. (dir.) 2002. *Les valeurs au Luxembourg. Portrait d'une société au tournant du 3e millénaire*. Luxembourg: Saint-Paul.

Leresche J.-P., Joye D., Bassand M. 1995. *Métropolisations: interdépendances mondiales et implications lémaniques*. Geneva: Georg.

Leresche J.-P., Saez G. 2002. Political frontier regimes: towards cross-border governance? In Perkmann M., Sum N.-L. (eds.) *Globalization, regionalization and cross-border regions*. London: Palgrave MacMillan, 77-99.

Lévy J. 1995. Paris métropolitains. Réseaux et territoires dans l'espace parisien. In Leresche J.-P., Joye D., Bassand M. (eds.) *Métropolisations: interdépendances mondiales et implications lémaniques*. Geneva: Georg, 57-73.

Lévy J. 1997. Penser la ville: un impératif sous toutes les latitudes. *Cahiers d'Etudes sur la Méditerranée Orientale et le Monde Turco-iranien*, 24: 25-38.

Lévy J. 2003. Capital spatial. In Lévy J., Lussault M. (eds.) *Dictionnaire de la géographie et de l'espace des sociétés*. Paris: Belin, 124-126.

Lévy J. 2006. Les espaces de l'âge au risque de l'historicité. In Brunel P., Attias-Donfut C., Morval J., Lévy J. (eds.) *Penser l'espace pour lire la vieillesse*. Paris: PUF, 71-96.

Lévy J., Lussault M. (eds.) 2003. *Dictionnaire de la géographie et de l'espace des sociétés*. Paris: Belin.

Ley D. 1996. *The new middle class and the remaking of the central city*. Oxford: Oxford University Press.

Lo V. 2003. Local codes and global networks. Knowledge access as a location factor in the financial industry. In Lo V., Schamp E.W. (eds.) *Knowledge, learning, and regional development*. Münster: Lit, 61-82.

Longcore T.R., Rees P.W. 1996. Information technology and downtown restructuring: the case of New York City's financial district. *Urban Geography*, 17(4): 354-372.

Lord S., Gerber P. 2009. Immigration et intégration: trajectoires résidentielles (inter)nationales et dynamiques ségrégatives locales au Luxembourg. *Espace, Populations, Sociétés*, 1: 85-103.

Lösch A. 1940. *Die räumliche Ordnung der Wirtschaft*. Jena: Gustav Fischer.

Lotscher L. 1991. Die oberrheinische Regio-Kooperation. Ein Modell fur Europa? *Geographische Rundschau*, 43(9): 520-526

Lüthi S., Thierstein A., Goebel V. 2008. Intra-firm and extra-firm linkages of the knowledge economy. The case of the mega-city region of Munich. *GaWC Research Bulletin*, 269. http://www.lboro.ac.uk/gawc/rb/rb269.html.

Luxembourg for Finance. 2009. *The Luxembourg financial centre*. http://www.lff.lu/key-facts.

Malmberg A., Maskell P. 2002. The elusive concept of localization economies: towards a knowledge-based theory of spatial clustering. *Environment and Planning A*, 34: 429-449.

Markusen A. 1996. Sticky places in slippery space: a typology of industrial districts. *Economic Geography*, 72(3): 293-313.

Martin R. 1999. The new economic geography of money. In Martin M. (ed.) *Money and the space economy*. Chichester: John Wiley, 3-27.

Martin R., Sunley P. 2003. Deconstructing clusters: chaotic concept or policy panacea? *Journal of Economic Geography*, 3: 5-35.

Martinelli F. 1992. Services aux producteurs et développement régional. *Espaces et Sociétés*, 66-67: 185-216.

Martinez O.J. 1994. The dynamics of border interaction. New approaches to border analysis. In Schofield C. (ed.) *Global boundaries. World boundaries*. London: Routledge, 1-15.

Maurin E. 2002. *L'égalité des possibles: la nouvelle société française*. Paris: Seuil.

Maurin E. 2004. *Le ghetto français: enquête sur le séparatisme social*. Paris: Seuil.

McDowell L. 1997. A tale of two cities? Embedded organizations and embodied workers in the City of London. In Lee R., Willis F. (eds.) *Geographies of economies*. London: Arnold, 118-129.

Meijers E. 2005. Polycentric uyrban regions and the quest for synergy: is a network of cities more than the sum of the parts? *Urban Studies*, 42(4): 765-781.

Merlin P., Choay F. 2000. *Dictionnaire de l'urbanisme et de l'aménagement.* Paris: PUF.

Ministère de l'Intérieur et de l'Aménagement du territoire (MIAT) 2003. *Programme directeur d'aménagement du territoire*. Luxembourg.

Mission opérationnelle transfrontalière (MOT) 2006. *Bonnes pratiques de gouvernance dans les agglomérations transfrontalières en Europe*. Report to the French-Belgian study group. Paris: MOT.

Mondada L. 2000. *Décrire la ville. La construction des savoirs urbains dans l'interaction et dans le texte*. Paris: Anthropos.

Motte A. 2007. *La notion de planification stratégique spatialisée (Strategic spatial Planning) en Europe (1995-2005)*. La Défense: PUCA.

Musterd S. 2005. Social and ethnic segregation in Europe: levels, causes, and effects. *Journal of Urban Affairs*, 27(3): 331-348.

Musterd S., Ostendorf W. (eds.) 1998. *Urban segregation and the welfare state: inequality and exclusion in western cities*. London, New York: Routledge.

Newling B.E. 1966. The spatial variation of urban population densities. *Geographical Review*, 59(2): 242-252.

Newman P., Kenworthy J.R. 1999. *Sustainability and cities: overcoming automobile dependence*. Washington, DC: Island Press.

Newman P., Thornley A. 1996. *Urban planning in Europe: international competition, national systems and planning projects*. London: Routledge.

Nielsen T.A.S., Hovgesen H.H. 2008. Exploratory mapping of commuter flows in England and Wales. *Journal of Transport Geography*, 16: 90-99.

North D.A. 1990. *Institutions, institutional change and economic performance*. Cambridge: Cambridge University Press.

Nottrot J. 1985. Luxemburg. Beiträge zur Stadtgeographie einer europäischen Hauptstadt und eines internationalen Finanzplatzes. *Innsbrucker Geographische Studien*, 12.

O'Dowd L. 2002. Transnational integration and cross-border regions in the European Union. In Anderson J. (ed.) *Transnational democracy: political spaces and border-crossings*. London: Routledge, 111-128.

O'Dowd L. 2003. The changing significance of European borders. In Anderson J., O'Dowd L., Wilson T.M. (eds.) *New borders for a changing Europe: cross-border cooperation and governance*. London: Frank Cass, 13-36.

OCSTAT 2009. *Frontaliers du canton de Genève en 2008*. Geneva, Cantonal Statistical Office. http://www.ge.ch/statistique/tel/publications/2009/resultats/dg-rs-2009-04.pdf.

OECD 2006a. *Innovation and knowledge-intensive service activities*. Paris: OECD.

OECD 2006b. *Competitive cities in the global economy*. Paris: OECD Territorial reviews.

OECD 2008. *Luxembourg 2008*. Paris: OECD Economic Surveys.

Ohmae K. 1995. *The end of the nation-state: the rise of regional economies*. New York: The Free Press.

Orfeuil J.-P. 2000. *L'évolution de la mobilité quotidienne. Comprendre les dynamiques, éclairer les controverses*. Arcueil: INRETS.

Paasi A. 1986. The institutionalization of regions: a theoretical framework for understanding the emergence of regions and the constitution of regional identity. *Fennia*, 164: 104-46.

Paasi A. 1991. Deconstructing regions: notes on the scales of spatial life. *Environment and Planning A*, 23(2): 239-256.

Painter J. 2009. Territoire et réseau: une fausse dichotomie? In Vanier M. (ed.) *Territoires, territorialité, territorialisation. Controverses et perspectives*. Rennes: Presses universitaires de Rennes.

Palsky G. 1996. *Des chiffres et des cartes. Naissance et développement de la cartographie quantitative au XIX*e *siècle*. Paris: Comité des travaux historiques et scientifiques.

Paltridge B. 2006. *Discourse analysis*. London, New York: Continuum.

Pandit N.R., Cook G.A.S. 2003. The benefits of industrial clustering: insights from the British financial services industry at three locations. *Journal of Financial Services Marketing*, 7(3): 230-245.

Pandit N.R., Cook G.A.S., Swann P.G.M. 2001. The dynamics of industrial clustering in British financial services. *The Service Industries Journal*, 21(4): 33-61.

Pansch H. 2009. Vom Findel in die Welt – Die Entwicklung des Flughafens. In Bousch P., Chilla T., Gerber P., Klein O., Schulz C., Sohn C., Wiktorin D. (eds.) *Atlas du Luxembourg*. Cologne: Emons, 118-119.

Park R.E., Burgess E.W., McKenzie R.D. 1925. *The city*. Chicago: University of Chicago Press.

Parr J.B. 1999a. Growth-pole strategies in regional economic planning: retrospective view. Part 1. Origins and advocacy. *Urban Studies*, 36(7): 1195-1215.

Parr J.B. 1999b. Growth-pole strategies in regional economic planning: retrospective view. Part 2. Implementation and outcomes. *Urban Studies*, 36(8): 1247-1268.

Pastorel J.P. 2007. Cités et reterritorialisation: la France à l'heure des grandes métropoles. *Victoria University of Wellington Law Review*, 38(1), 35-50.

Peach C. 2001. *The ghetto and the ethnic enclave*. International Seminar on Segregation in the City, July 26-28, Lincoln Institute of Land Policy, Cambridge (MA).

Peck J. 2005. Economic sociologies in space. *Economic Geography*, 81(2): 129-175.

Perkmann M. 2003. Cross-border regions in Europe: significance and drivers of regional cross-border co-operation. *European Urban and Regional Studies*, 10(2): 153-171.

Perkmann M. 2007. Policy entrepreneurship and multilevel governance: a comparative study of European cross-border regions. *Environment and Planning C*, 25: 861-879.

Perkmann M., Sum N.-L. (eds.) 2002. *Globalization, regionalization and cross-border regions*. New York: Palgrave Macmillan.

Philippe J., Léo P.-Y., Boulianne L.-M. (eds.) 1998. *Services et métropoles. Formes urbaines et changement économique*. Paris: L'Harmattan.

Phillips L., Jorgensen M.W. 2002. *Discourse analysis as theory and method.* London: Sage.

Pieretti P., Bourgain A. 2003. Technological spillovers of financial centers on a regional economy: a macroeconomic approach applied to Luxembourg. In Thierstein A., Schamp E.W. (eds.) *Innovation, finance, and space*. Frankfurt: Frankfurt Studies on Economic and Social Geography, 19-27.

Pieretti P., Bourgain A., Courtin P. 2007. *Place financière de Luxembourg. Analyse des sources de ses avantages compétitifs et de sa dynamique.* Brussels: De Boeck.

Pigeron-Piroth I. 2008. La mobilité résidentielle des personnes travaillant au Luxembourg (1995-2005). *Bulletin du STATEC*, 1. Luxembourg: STATEC.

Pinol J.-L. 1992. *Le monde des villes au XIXe siècle*. Paris: Hachette.

Piore M., Sabel F. 1984. *The second industrial divide: possibilities for prosperity*. New York: Basic Books.

Polanyi M. 1967. *The tacit dimension*. New York: Anchor Books.

Poon J.P.H. 2003. Hierarchical tendencies of capital markets among international financial centers. *Growth and Change*, 34(2): 135-156.

Poon J.P.H., Eldredge B., Yeung D. 2004. Rank size distribution of international financial centers. *International Regional Science Review*, 27(4): 411-430.

Porteous D. 1999. The development of financial centres: location, information externalities and path dependency. In Martin M. (ed.) *Money and the space economy*. Chichester: John Wiley, 95-114.

Pries L. 2001. The approach of transnational social spaces: responding to new configurations of the social and the spatial. In Pries L. (ed.) *New transnational social spaces*. London: Routledge, 3-33.

Pumain D. 1993. *Villes, métropoles, régions urbaines... Un essai de clarification des concepts*. Paper presented at the conference "Métropoles et aménagement du territoire", Institut d'aménagement et d'urbanisme de la Région Île de France, 12-13 May.

Pumain D., Offner J.M. 1996. Réseaux de villes. In Pumain D., Offner J.M. (eds.) *Réseaux et territoires. Significations croisées*. La Tour d'Aigues: Aube.

Pumain D., Paquot T., Kleinschmager R. 2006. *Dictionnaire. La ville et l'urbain*. Paris: Economica.

Quadagno J. 1999. *Aging and the life course.* New York: McGraw-Hill.

Ratti R. 1994. Spatial effects of frontiers: overview of different approaches and theories of border region development. In Nijkamp P. (ed.) *New borders and old barriers in spatial development*. Aldershot: Averbury, 15-33.

Ratti R., Reichman S. (eds.) 1993. *Theory and practice of transborder cooperation*. Basel, Frankfurt am Main: Helbing & Lichtenhahn.

Reitel B. 2006. Governance in cross-border agglomerations in Europe. The examples of Basel and Strasbourg. *Europa Regional*, 14(1): 9-21.

Reitel B. 2007. Les agglomérations transfrontalières: des systèmes urbains en voie d'intégration? Les espaces urbains de la "frontière" du territoire français. *Geographica Helvetica*, 1(7): 5-15.

Reitel B., Zander P., Piermay J.-L, Renard J.-P. (eds.) 2002. *Villes et frontières*. Paris: Anthropos.

Reynaud A. 1981. *Société, espace et justice*. Paris: PUF.

Roberts R. 2008. *The City. A guide to London's global financial centre*. London: The Economist.

Roberts S. 1994. Fictitious capital, fictitious spaces: the geography of offshore financial flows. In Corbridge S., Thrift N., Martin R. (eds.) *Money, power and space*. Oxford: Blackwell, 91-115.

Roux E., Vanier M. 2008. *La périurbanisation: problématiques et perspectives*. Paris: DIACT.

Rozenblat C., Cicille P. 2003. *Les villes européennes. Analyse comparative*. Montpellier: Maison de la géographie.

Rozenblat C., Pumain D. 2007. Firm linkages, innovation and the evolution of urban systems. In Taylor P., Derudder B., Saey P., Witlox F. (eds.) *Cities in globalization. Practices, policies and theories*. London: Routledge.

Saez G., Bassand M. 1997. Les recompositions de l'action publique en contexte métropolitain et transfrontalier. In Saez G., Leresche J.-P., Bassand M. (eds.) *Gouvernance métropolitaine et transfrontalière*. Paris: L'Harmattan, 9-46.

Saez G., Leresche J.-P., Bassand M. (eds.) 1997. *Gouvernance métropolitaine et transfrontalière*. Paris: L'Harmattan.

Sassen S. 2001. *The global city. New York, London, Tokyo*. Princeton: Princeton University Press.

Saurugger S. 2003. *Européaniser les intérêts. Les groupes d'intérêt économiques et l'élargissement de l'Union européenne*. Paris: L'Harmattan.

Schamp E.W. 2009. Das Finanzzentrum – ein Cluster? Ein multiskalarer Ansatz und seine Evidenz am Beispiel von Frankfurt/Rhein-Main. *Zeitschrift für Wirtschaftsgeographie*, 53(1-2): 89-105.

Schmidt S. 2009. Land use planning tools and institutional change in Germany: recent developments in local and regional planning. *European Planning Studies*, 17: 1907-1921.

Schneider M., Pigeron-Piroth I. 2009. Une aire résidentielle transfrontalière en expansion. In Bousch P., Chilla T., Gerber P., Klein O., Schulz C., Sohn C., Wiktorin D. (eds.) *Atlas du Luxembourg*. Cologne: Emons, 180-182.

Schuler M., Dessemontet P., Jemelin C., Jarne A., Pasche N., Haug W. 2007. *Atlas des mutations spatiales de la Suisse*. Zurich: NZZ.

Scott A.J. 1998. *Regions and the world economy. The coming shape of global production, competition, and political order*. Oxford: Oxford University Press.

Scott A.J., 2001. Globalization and the rise of city-regions. *European Planning Studies*, 9: 813-826.

Scott J.W. 1989. Transborder cooperation, regional initiatives and sovereignty conflicts in Western Europe: the case of the Upper Rhine valley. *Publius: The Journal of Federalism*, 19(1): 139-156.

Scott J.W. 1999. European and North American contexts for cross-border regionalism. *Regional Studies*, 33(7): 605-17.

Scott J.W. 2002. A networked space of meaning? Spatial politics geostrategies of European integration. *Space and Polity*, 6(2): 147-167.

Sélimanovski C. 2008. *La frontière de la pauvreté*. Rennes: Presses universitaires de Rennes.

Shaw D., Sykes O. 2004. The concept of polycentricity in European spatial planning: reflections on its interpretation and application in the practice of spatial planning. *International Planning Studies*, 9(4): 283-306.

Shen J. 2004. Cross-border urban governance in Hong Kong: the role of state in a globalizing city-region. *The Professional Geographer*, 56(4): 530-543.

Sohn C. 2006. Villes et agglomérations au Grand-Duché de Luxembourg, Proposition d'une nomenclature des unités urbaines. *Population & Territoire*, 10, CEPS/INSTEAD.

Sohn C. 2009a. Des villes entre coopération et concurrence. Analyse des relations culturelles transfrontalières dans le cadre de Luxembourg et Grande Région, Capitale européenne de la culture 2007. *Annales de géographie*, 667: 228-246.

Sohn C. 2009b. Géographie des activités métropolitaines à Luxembourg-Ville. In Bousch P., Chilla T., Gerber P., Klein O., Schulz C., Sohn C., Wiktorin D. (eds.) *Atlas du Luxembourg*. Cologne: Emons, 56-59.

Sohn C. 2010. El papel ambivalente de las fronteras en la construcción de la metrópolis transfronterizas en Europa. Los casos de Basilea, Ginebra y Luxemburgo. *Documents d'Anàlisi Geogràfica*, 56(1): 167-184.

Sohn C., Reitel B., Walther O. 2009. Cross-border metropolitan integration in Europe: the case of Luxembourg, Basel and Geneva. *Environment & Planning C: Government and Policy*, 27: 922-939.

Sohn C., Walther O. 2008. Géographie des activités économiques métropolitaines au Luxembourg. *Population & Territoire*, 14, CEPS/INSTEAD.

Sohn C., Walther O. 2009. Métropolisation et intégration transfrontalière: le paradoxe luxembourgeois. *Espaces & Sociétés*, 138: 51-67.

Sokol M., van Egeraat C. 2005. *Locational strategies of advanced producer services in Dublin: towards a polycentric mega-city region?* Paper presented at the Regional Studies Association International Conference, University of Aalborg, 28-31 May.

STATEC 2009a. La société luxembourgeoises depuis le milieu du XIXe siècle dans une perspective économique et sociale. *Cahier économique*, 108. Luxembourg: STATEC.

STATEC 2009b. *The Luxembourg economy. A kaleidoscope*. Luxembourg: STATEC.

STATEC 2009c. *Annuaire statistique du Luxembourg*. Luxembourg: STATEC.

Storper M., Venables A.J. 2004. Buzz: face-to-face contact and the urban economy. *Journal of Economic Geography*, 4(4): 351-370.

Strassoldo R. 1970. *From barrier to junction: towards a sociological theory of borders*. Gorizia: ISIG.

Strüver A. 2004. We are only allowed to re-act, not to act': Eurocrats' strategies and bordelanders' tactics in a Dutch-German cross-border region. In Kramsch O., Hooper B. (eds.) *Cross-border governance in the European Union*. London: Routlege, 25-40.

Tabourin E., Andan O., Routhier J.-L. 1995. *Les formes de la croissance urbaine. Le modèle de René Buissière appliqué à l'agglomération lyonnaise*. LET Research Report, Lyon.

Taylor D.R.F. 1983. Graphic communication and design in contemporary cartography: an introduction. In Taylor D.R.F. (ed.) *Graphic communication and design in contemporary cartography*. Chichester: John Wiley & Sons, 1-10.

Taylor P.J. 2000. World cities and territorial states under conditions of contemporary globalization. *Political Geography*, 19: 5-32.

Taylor P.J. 2004. *World city network. A global urban analysis*. London: Routledge.

Taylor P.J., Beaverstock J., Cook G., Pandit N., Pain K., Greenswood H. 2003. *Financial services clustering and its significance for London*. Full Report. London: Corporation of London.

Taylor P.J., Walker D.R.F., Beaverstock J.V. 2002. Firms and their global service networks. In Sassen S. (ed.) *Global networks. Linked cities*. New York: Routledge, 93-115.

The Economist. 2007. *Special report on financial centres*, September 13th.

Thierstein A., Lüthi S., Kruse C., Gabi S., Glanzmann L. 2008. Changing value chain of the Swiss knowledge economy: spatial impact of intra-firm and inter-firm networks within the emerging mega-city region of Northern Switzerland. *Regional Studies*, 42(8): 1113-1131.

Thisse J.-F., van Ypersele T. 1999. Métropoles et concurrence territoriale. *Economie et Statistique*, 326: 19-30.

Thrift N. 1994. On the social and cultural determinants of international financial centres: the case of London. In Corbridge S., Thrift N., Martin R. (eds.) *Money, power and space*. Oxford: Blackwell, 327-355.

Tobler W.R. 1976. Spatial interaction patterns. *Journal of Environmental Systems*, 6(4): 271-301.

Tobler W.R. 1978. Migration fields. In Clark W., Moore E. (eds.) *Population mobility and residential change*, Studies in Geography, 25, 215-232.

Tobler W.R. 1981. A model of geographical movement. *Geographical Analysis*, 13(1): 1-20.

Tobler W.R. 1982. *Cartographic study of movement tables*. Proceedings. National Computer Graphics Association, Anaheim.

Trausch G. 1994. *La Ville de Luxembourg. Du château des comtes à la métropole européenne*. Anvers: Fonds Mercator Paribas.

Trausch G. 2003. *Histoire du Luxembourg. Le destin européen d'un « petit pays »*. Toulouse: Privat.

Tschoegl A.E. 2000. International banking centers, geography, and foreign banks. *Financial Markets, Institutions & Instruments*, 9(1): 1-32.

UK National Statistics. 2007. *Focus on London*. London: Palgrave Macmillan.

Urry J. 2003. *Global complexity*. Cambridge: Polity.

Van der Velde M., van Houtum H. (eds.) 2000. *Borders, border regions and people*. London: Pion.

Van Geenhuizen M., Ratti R. (eds.) 2001. *Gaining advantage from open borders. An active space approach to regional development*. Aldershot: Ashgate.

Van Grunsven L. 1995. Industrial regionalization and urban-regional transformation in Southeast Asia: the Sijori growth triangle considered. *Malaysian Journal of Tropical Geography*, 26(1): 47-65.

Van Houtum H. 2000. An overview of European geographical research on borders and border regions. *Journal of Borderlands Studies*, 15(1): 57-83.

Van Houtum H., Gielis R. 2006. Elastic migration: the case of Dutch short-distance transmigrants in Belgian and German borderlands. *Tijddschrift voor Economische en Sociale Geografie*, 97(2): 195-202.

Van Winden W., van den Berg L., Pol P. 2007. European cities in the knowledge economy: towards a typology. *Urban Studies*, 44(3): 525-549.

Vandermotten C., 2010. Les aires urbaines transfrontalières en Europe. *Mosella*, 32(1-4): 13-23.

Varbanova L. 2008. *Luxembourg and Greater Region, European capital of culture 2007.* Final Report. Luxembourg.

Väyrynen R. 2003. Regionalism: old and new. *International Studies review*, 5: 25-51.

Veltz P. 1996. *Mondialisation, villes et territoire. L'économie d'archipel.* La Tour-d'Aigues: Aube.

Vignal C. 2006. Concilier mobilité résidentielle et mobilité professionnelle: cadrage statistique et pistes de réflexion. *Horizons stratégiques*, 2(2): 37-47.

Walther O., Dautel V. 2010. Intra-regional employment growth in Luxembourg (1994-2005). *Geografiska Annaler: Series B, Human Geography*, 92(1): 45-63.

Walther O., Schulz C. 2009. Finanzplatz Luxemburg. Vom "Steuerparadies" zur Investmentfonds-Kapitale. *Geographische Rundschau*, 61(1): 30-35.

Webber M. 1964. The urban place and the nonplace urban realm. In Webber M., Dyckman D., Foley D., Guttenberg A., Wheaton W., Wurster C.B. (eds.) *Explorations in urban structure*. Philadelphia: University of Pennsylvania Press, 79-137.

Weber A. 2008. Aménagement urbain et post-keynésianisme. *Métropoles*, 4. http://metropoles.revues.org/document3452.html.

Weiss A. 2005. The transnationalization of social inequality: conceptualizing social positions on a world scale. *Current Sociology*, 53(4): 707-728.

Wiel M. 1999. *La transition urbaine ou le passage de la ville-pédestre à la ville-motorisée*. Wavre: Mardaga.

Yang C. 2005. An emerging cross-boundary metropolis in China. Hong Kong and Shenzhen under "Two Systems". *International Development Planning Review*, 27(2): 195-225.

Yang C. 2006. The geopolitics of cross-boundary governance in the Greater Pearl River Delta, China: a case study of the proposed Hong Kong-Zhuhai-Macao bridge. *Political Geography*, 25: 817-835.

Yapa L., Polese M., Wolpert J. 1971. Interdependencies of commuting, migration and job location. *Economic Geography*, 47(1): 59-72.

Zanardelli M., Brosius J. 2009. Les recrutements au Luxembourg. *Population & Emploi*, 14, CEPS/INSTEAD.

Zaninetti J.-M. 2005. *Statistique spatiale: méthodes et applications géométriques*. Paris: Lavoisier.

Zipf G.K. 1949. *Human behaviour and the principle of least effort: an introduction to human ecology.* Cambridge: Addison-Wesley.

Notes on Contributors

Eric Auburtin is a teacher of geography and geopolitics in Nancy (Lycée Henri Poincaré) and Lyon (École normale supérieure). He is also associated with the French Institute of Geopolitics at the University of Paris 8. As a geopolitical geographer, his research interests encompass the analysis of cross-border spaces in northwestern Europe. He is the co-author (with B. Giblin) of *Nouvelle géopolitique des régions françaises*.

Ralf Bläser is a geographer. His main research focus is the geography of civil society. He currently works for the German Council of Science and Humanities (Wissenschaftsrat).

Samuel Carpentier is an urban geographer. He is a researcher at CEPS/INSTEAD and associate researcher at the University of Strasbourg (ERL LIVE). In his research, Samuel is investigating the interactions between daily and residential mobility. He is also the editor in chief of *Articulo – Journal of Urban Research.*

Franz Clément has been a researcher in industrial relations at CEPS/INSTEAD since 1996. He is also affiliated to the *Conservatoire National des Arts et Métiers* in Paris. His research entails two thematics: industrial relations in Luxembourg and in the Greater Region.

Antoine Decoville is an urban planner/geographer. As a researcher at CEPS/INSTEAD, he is investigating the urban and regional development of Luxembourg, as well as governance and spatial planning issues. He is also in charge of the Spatial Planning Observatory, a tool for decision support for Luxembourg's Ministry of Sustainable Development and Infrastructures.

Michaela Gensheimer is working as cluster coordinator/project expert at the ESPON Coordination Unit, the secretariat of the ESPON 2013 Programme. She is developing and following European research projects that provide evidence on territorial development opportunities and dynamics in order to support policy development for the European territory. Before joining the ESPON CU in 2007, Michaela worked as researcher at the University of Luxembourg, contributing to the METROLUX project.

Philippe Gerber has been researcher in the field of urban and social geography at CEPS/INSTEAD since 2001. He is also associate lecturer at the University of Luxembourg and associate researcher at the University of Strasbourg (ERL LIVE). Philippe is coordinating several research projects that focus on the interactions between daily and residential mobility.

Olivier Klein is a researcher at CEPS/INSTEAD. With a specialisation in geovisualisation, he conducts research on daily mobilities and their cartographic representations. His work also includes, more generally, theory and design of appropriate tools for the dissemination of geographic information in order to observe the territory and its development.

Christian Lamour is an urban planner/geographer. He is a researcher at CEPS/INSTEAD, working as a project manager of cross-border cooperation in the Greater Region and as an expert in European city management projects. His main interests are the development of institutional and informal multi-level partnerships within cross-border conurbations and the relations between decision-making networks and urban communities.

Sébastien Lord is a researcher in urban studies at CEPS/INSTEAD. He holds a PhD in regional planning and a master's degree in Architectural Science from *Université Laval* (Québec, Canada). His research focuses on housing, aging, daily and residential mobilities, immigration, socio-spatial inequalities and urban forms.

Bernard Reitel is a senior lecturer at the University of Upper Alsace in Mulhouse where he teaches urban and political geography. In his research, Bernard is investigating transformations in urban planning in border cities like Basel, Berlin and Strasbourg and the evolution of governance systems in border regions in Western Europe. He is the editor (with Patricia Zander, Jean-Luc Piermay and Jean-Pierre Renard, 2002) of *Villes et frontières*.

Christian Schulz is professor of European Sustainable Spatial Development and Analysis at the University of Luxembourg. As head of the Research Unit IPSE (Identités, Politiques, Sociétés, Espaces), his interests encompass spatial development in cross-border regions, research on industrial clusters, as well as environmental economic geography.

Christophe Sohn is an urban geographer. He is a researcher at CEPS/INSTEAD and associate assistant professor at the University of Luxembourg. In his research, Christophe is investigating the spatial and political construction of cross-border metropolitan regions as privileged sites of globalisation. He is currently leading a project focusing on cross-border governance and policy networks in four European cross-border regions.

Olivier Walther is a researcher at the CEPS/INSTEAD and the co-editor of *Articulo – Journal of Urban Research*. As an economic geographer, his research interests range from cross-border economic networks in Europe and West Africa to the geography of international finance. He is the author of *Affaires de patrons. Villes et commerce transfrontalier au Sahel* (2008).

Peter Lang – The website
Discover the general website of the Peter Lang publishing group:
www.peterlang.com